Railways

Volume I

Studies in Transport History

Series Editor: John Armstrong

Coastal and Short Sea Shipping
John Armstrong

Railways (volume I)
Terry Gourvish

Railways (volume II)
Geoffrey Channon

Road Transport in the Horse-drawn Era
Dorian Gerhold

Air Transport
Peter J. Lyth

Canals and Inland Navigation
Gerald Crompton

Motor Transport
Theo Barker

The World of Shipping
David M. Williams

Railways
Volume I

Edited by

TERRY GOURVISH

© The publishers of *The Journal of Transport History*, 1996;
Editor's introduction © Terry Gourvish, 1996

All rights reserved. No part of this publication may be reproduced, stored in a retrieval system, or transmitted in any form or by any means electronic, mechanical or photocopying, recording, or otherwise without the prior permission of the publisher.

Published by
SCOLAR PRESS
Gower House
Croft Road
Aldershot
Hants GU11 3HR
England

Ashgate Publishing Company
Old Post Road
Brookfield
Vermont 05036–9704
USA

British Library Cataloguing in Publication Data

Railways, vol. 1.
  (Studies in Transport History)
  1. Railroads—History—19th century. 2. Railroads—History—20th century.
  I. Gourvish, T. R. (Terence Richard), 1943–
  385'.09

  ISBN 1–85928–299–7

Library of Congress Cataloging-in-Publication Data

Railways, vol. 1.
    p. cm. (Studies in Transport History). Vol. 2 edited by Geoffrey Channon.
  ISBN 1–85928–299–7 (cloth)
  1. Railroads—History. I. Gourvish, T. R. (Terence Richard).
II. Channon, Geoffrey. III. Series.
HE1021.R35   1996
385'.09—dc20   96–9261
                                                                     CIP
ISBN 1 85928 299 7
Printed on acid free paper
Printed and bound in Great Britain by Ipswich Book Co. Ltd., Ipswich, Suffolk

Contents

Series Editor's Preface   vii

Introduction: railways in the macro-economy   ix
    Terry Gourvish

1   Railways and economic growth: a review article   1
    2nd series, vol. I, no. 4 (1972), pp. 238-49
    D. H. Aldcroft

2   Railways and the Scottish transport system in the nineteenth century   13
    2nd series, vol. I, no. 3 (1972), pp. 133-45
    W. Vamplew

3   Railroads and the American economy: the Fogel thesis in retrospect   26
    3rd series, vol. IV, no. 2 (1983), pp. 20-34
    D. L. Lightner

4   Railways and land values   41
    3rd series, vol. IV, no. 2 (1983), pp. 35-49
    J. J. Pincus

5   Private enterprise or public utility? Output, pricing and investment in English and Welsh railways, 1879-1914   57
    3rd series, vol. I, no. 1 (1980), pp. 9-28
    P. J. Cain

6   The capitalisation of Britain's railways, 1830-1914   77
    3rd series, vol. V, no 1 (1984), pp. 1-24
    R. J. Irving

7   The state and Indian railway performance, 1870-1920, Part 1: Financial efficiency and standards of service   101
    3rd series, vol. II, no. 2 (1981), pp. 1-15
    R. O. Christensen

	CONTENTS	
8	The state and Indian railway performance, 1870-1920, Part 2: The government, rating policy and capital funding 3rd series, vol. III, no. 1 (1982), pp. 21-34 *R. O. Christensen*	117
9	The German National Railway between the world wars: modernisation or preparation for war? 3rd series, vol. XI, no. 1 (1990), pp. 40-60 *A. C. Mierzejewski*	131
10	A. D. Chandler's 'visible hand' in transport history: a review article 3rd series, vol. II, no. 1 (1981), pp. 53-64 *G. Channon*	153
11	Myth and rationality in management decision-making: the evolution of American railroad product costing, 1870-1970 3rd series, vol. XII, no. 1 (1991), pp. 1-10 *G. L. Thompson*	165

Studies in Transport History

Series Editor's Preface

The idea for this series originated in a meeting of the Editorial Board of *The Journal of Transport History*. As part of the celebrations to mark the fortieth anniversary of the founding of the journal, a classified index of the articles appearing in each of the series was compiled by Julie Stevenson and published in the September 1993 number of the journal. This exercise revealed the wealth of material that lay in the journal but which was largely forgotten and relatively inaccessible, unless a full run of the journal was close to hand. The index took the first step in making it simpler to determine what was in the journal. The Editorial Board decided it should continue this process by making the best of the essays more readily available.

Hence it is proposed to issue a series of volumes, each containing a number of articles reprinted from *The Journal of Transport History*, dealing with a particular mode of transport. As initially planned there are eight volumes in this series, each covering one form of transport and edited by a different member of the Editorial Board. Each volume contains about ten articles, reprinted from *The Journal of Transport History*. The articles have been selected because they were seminal at the date of their publication and have stood the test of time. Hence they are still important contributions to the state of our knowledge on the topic.

Each volume also contains an introduction by the editor of the volume, which seeks to contextualise the articles selected; that is, to show how each contribution fitted into the state of knowledge at the time of its publication, to explain its importance, to indicate any more-recent literature on the topic that might moderate its findings, and suggest how the debate has moved since the initial publication. Each article retains its original pagination, as in the journal. This will allow citations to be made to the original source.

Thus this series aims to make easily available the most important essays which have appeared in *The Journal of Transport History* over the last forty years, collected together on a thematic basis. It is hoped that this will encourage and provoke further research into transport history, in order to carry the debates into the twenty-first century.

John Armstrong
1996

Thames Valley University

Introduction: railways in the macro-economy

TERRY GOURVISH

Writing about railways has always had an eclectic character, and academic writing about the industry reflects this deep-seated eclecticism.[1] From the start, in the first half of the nineteenth century, some writers focused on the speculative fever associated with the promotion and construction of the great railway companies; others were obsessed with the technological achievement, beginning with the great engineering works and the resplendent steam machines; others sought to measure the economic and social benefits; yet others warned of the social and economic costs. Historians, economists, politicians, engineers, journalists, enthusiastic amateurs, all wished to have their say; and this tradition has persisted. In relation to this volume of collected articles, which focuses in the main on *macro*-economic subjects,[2] we should note that almost from the beginning there were commentators who embraced a methodological and a macro-economic concern. Thus, Dionysius Lardner, in his insightful if idiosyncratic *Railway Economy* of 1850, analysed the economic, social and engineering consequences of the early railway period, with chapters on management structures, relations with government and comparisons with railways in other countries.[3] Sixteen years later Dudley Baxter produced the first 'social saving' calculation of the benefits which railways had brought to the British economy when he asserted that 'had the railway traffic of 1865 been conveyed by canal and road at the pre-railway rates, it would have cost three times as much'.[4]

The articles selected from *The Journal of Transport History* in this volume reflect this interest and, in particular, the 'high point' of concern about the linkage between railways and economic growth, which began in 1964 with Brian Mitchell's seminal article on railways and British economic growth, and in the United States, with Robert Fogel's path-breaking monograph on the implications for growth of the American railroads.[5] Fishlow's wider-ranging contribution to the American debate followed in 1965,[6] but the scholarship on Britain had to wait for the publications of two young research students, Wray Vamplew and Gary Hawke, who were influenced by, and reacted very differently to, the considered calculations of Mitchell and the more adventurous, 'social savings' approach adopted by the American economists Fogel and Fishlow.

In 1970 Gary Hawke's path-breaking monograph, *Railways and Economic Growth in England and Wales,* heralded a decade of academic interest in the search for more precise numbers with which to measure the railways' contribution to growth.[7] Patrick O'Brien published one of the most important volumes, a short textbook setting out the state of scholarship in the USA, UK, Italy and Russia.[8] This flurry of interest, summarised in Gourvish's pamphlet for the Economic History Society, published in 1980,[9] was followed by a period of more mature reflection in the 1980s. This saw attempts both to broaden the parameters of the debate and to pursue research in a more comparative context. In 1983 O'Brien edited an impressive collection of papers which yielded the results of new research on the role of railways in Belgium, France, Germany, Italy and Spain.[10] Some years later, in 1990, the French railway historian, François Caron, organised an 'A' session of the International Economic History Congress in Leuven, which made the railways, with their contribution to growth, a central concern.[11] Since then interest in this area has generally diminished. A good example is the perfunctory treatment of railways in the updated three-volume economic history of Britain, edited by Floud and McCloskey.[12] However, we have a significant exception in James Foreman-Peck, who is currently the most active British scholar in the field of railways and economic growth, though perhaps he is better known for his facility with numbers than for his detailed knowledge of the railway industry itself.[13] In his essay on 'Railways and late Victorian economic growth', published in 1990, he built upon the work of Hawke et al., producing new social savings estimates for railways in 1865, 1890 and 1910 which indicated that the savings were large. Foreman-Peck therefore concluded that not only were railways of considerable importance to the late Victorian economy, but issues regarding their efficiency and control were in consequence an important area of debate for scholars. This theme was developed in his article on 'Natural monopoly', published in 1987, and in 1994 in a book on *Public and Private Ownership of British Industry*, co-authored with Robert Millward. One of the principal arguments here was that free enterprise investment had made railway capital costs about 50 per cent higher than they might have been, representing a loss of national income of about 0.75 per cent in 1906.[14] The material selected in this volume reflects the broader historiography outlined above. Thus, no articles have been selected which predate 1972. Of the eleven contributions, two were published in 1972, when interest in the economic growth debate was at its height, but seven appeared in the period 1980-84, when the new editors, Gourvish and Turnbull, encouraged scholars to adopt a more methodological and analytical approach to railway history.

The articles which appeared in 1972 were important pieces of scholarship, and have been consulted by two generations of students. Derek Aldcroft not only provided a welcome review of Gary Hawke's seminal work; he produced one of the best critical assessments of the 'new economic history' approach to railways. He rightly pointed out that the techniques, while neither new nor particularly sophisticated in themselves, offered scholars a chance to resolve 'problems of specification in major areas of historical debate'.[15] After a short assessment of the significance of the work of Fogel, Fishlow, Conrad and Meyer, and Engerman, Aldcroft turned to Hawke's account. Praising it for the wealth and breadth of its coverage on early Victorian railways, he was more sceptical about its social savings measurement for 1865. Although it is difficult to trace the original source of critical comments about a particular work, it is in this article that we find most of the more commonly quoted criticisms of Hawke's calculation: the fragility of some of the theoretical assumptions, such as the constancy of canal costs over time; the limitations of the empirical evidence for key computations (e.g. canal costs); the method of comparing road and rail passenger services and fares; and the absence of a time savings calculation. Aldcroft has provided us with a major contribution to the subject which is still consulted widely.

A graduate student, Wray Vamplew, hearing that Gary Hawke was interested in the same subject, agreed to focus on the railways' role in Scotland, leaving England and Wales to Hawke. Here Vamplew's research led him to a much more agnostic position than his fellow student. He argued that it was difficult, if not impossible, to produce key measurements owing to the numerous pitfalls in attempting to quantify important variables. Thus, in a paper wittily entitled 'Nihilistic impressions of British railway history', he showed why he had abandoned the search for meaningful productivity estimates for Scottish railways over the period 1870-1900. Elsewhere he argued that even where measurement was feasible, 'inadequate data and the assumptions involved in quantification do not allow for absolute precision, and all we can hope to do is establish some order of magnitude'.[16]

Vamplew, unlike Hawke, did not publish a monograph based on his thesis. He preferred instead to disseminate his major findings in the journals. The article reproduced here was very much a supporting one to his main argument. In it he identified consumer benefits arising from the competition which the railways brought to the transport sector in Scotland. Vamplew also made the valid point that the existing alternative modes of transport were not destroyed. Although only five significant canals were still operational in Scotland by 1900, three of them were railway-owned and continued to provide a useful form of transport for freight. Furthermore, 'coastal shipping remained a formidable competitor of the railways throughout the nineteenth century', a theme taken up elsewhere in the Scolar collection.[17] However, for Vamplew's valid concerns about the data required to produce a social savings calculation for Scotland, and his commendable caution in approaching an estimate of the railways' contribution to Scottish economic development, the reader must look elsewhere.[18]

As representatives of the more theoretical literature on the railways in the 1980s we include two articles from the September 1983 issue of the *Journal*: David Lightner's retrospective on the Fogel thesis, and Jonathan Pincus's examination of the relationship between railway-building and land values. In the first, Lightner, an 'interested outsider' from Alberta, had the advantage of being able to draw on two decades of critical comment, including Fogel's more introspective, though still unrepentant, evaluation of the social saving approach in his article for the *Journal of Economic History* in 1979.[19] With the focus firmly on the American debate, Lightner concludes that the judgement on Fogel's work must be equivocal. His data and techniques are found to have been soundly based, with the upper bound estimate of the agricultural social saving 'better grounded than his critics had supposed'.[20] On the other hand, Fogel is held to remain weak in the manner in which an *overall* social saving calculation is then derived. Furthermore, Fogel's thesis can be challenged on utility grounds. 'Because of its conceptual weaknesses' argues Lightner, 'the Fogel thesis has not provided and cannot provide a definitive measurement of the contribution of railroads to American economic growth'.[21] This is a very useful and succinct summary of the debate.

Pincus makes a more novel contribution. He is concerned with what he calls 'the heart of the rural railway problem - the creation of land values'.[22] Conditions in Australia, where railways were publicly owned, are compared with those in the United States, where private railway companies were supported by government land grants. After a brief review of some of the literature on the two systems, Pincus provides what is essentially an essay in public choice economics using a simple model of railway economics. However, he also makes a contribution to the debate about public versus private ownership in the provision of rail transport. After evaluating the economic benefits and costs likely to accrue from the two systems, he points out a dilemma in public choice: 'how to internalise the marginal benefits generated by a utility like a railway in order that investment decision and pricing rules be economically efficient'.[23] The conclusion is that while there were defects with the US model it offered superior efficiency to the Australian model, where profits, including land rents, tended to be used to support an over-extended and heavily subsidised network.

Relevant to concerns about the impact of railways on economic growth are articles from two of our leading post-war railway historians, Peter Cain and Robert Irving. Both pieces involve a reworking of the key data used in the debate over the declining productivity of Britain's railways after 1870. Cain, contributing to the relaunch volume of the *Journal* in 1980, took up the challenge of calculating output figures. His new estimates of freight ton-mileage for England and Wales over the period 1871-1911 produced a substantial downward revision of Hawke's earlier figures for 1870-90. Hawke had ton-mileage rising by 110 per cent, from 5.0 billion in 1871 to 10.5 billion in 1890. Cain's estimates were based on a more circumspect (and realistic) appraisal of train loads and the average length of haul, and, in projecting backwards from a benchmark figure in 1920, due allowance was made for the strength of coastal shipping competition before 1914 and the rise in railway productivity during 1914-20. He thus came up with lower levels of output and a lower rate of increase of 83 per cent, from 4.2 billion to 7.7 billion.[24] These estimates then provoked a more general discussion of railway company performance. Output was found to have risen sharply in the 1870s, accompanied by falling freight rates. But in the 1880s output growth was more sluggish, while train loads stagnated and rates continued to fall. The rise in output for 1900-11 was impressive, but it was no greater than during the period 1880-1900; the difference lay in the substantial improvement in train loads after 1900, a response to falling rates at a time of overall inflation.

Cain's work added authority to those who had argued that the railways' operating productivity record after 1870 was disappointing and encouraged further work in this area. For example, Foreman-Peck produced new calculations indicating how poor the railways' total factor productivity growth was in comparison with that of the American railroads. The gain was apparently only about 1 per cent per annum for 1870-1914, half that achieved in the United States.[25] However, Cain threw his weight behind scholars such as Ashworth, Irving and Gourvish, who saw the industry's problems largely in terms of pressures beyond corporate control, and specifically the public service constraints demanded and secured by government and railway customers.[26]

Irving's article, on the other hand, challenged one of the major props of post-1870 historiography: the alleged over-capitalisation of the industry. In one of the most influential contributions of the last decade, he argued convincingly that the existing debate was off beam. For Irving, over-capitalisation was established much earlier, *before* the investment 'mania' of 1845-47. First, the Hawke and Reed data on capital investment, 1825-1912, were modified for the period 1870-90, by making full allowance for the extent of nominal additions to capital (watered capital) before 1890. The adjustment indicated that the contribution of loans and debentures to railway capital formation had been understated, and also highlighted the problems of Scottish railway companies from the early 1870s. Second, the conclusion was drawn that the real growth of railway capital from c.1875 to 1900 was more modest than had been suggested and the real rate of return on capital consequently higher.[27] Capital expenditure per mile (excluding nominal additions) was found to have doubled from £18,800 in 1830 to £37,400 in 1850; it then showed little change to 1870 (£36,900). Thereafter, capitalisation increased by 44 per cent to £53,000 in 1910, a rate of increase approximately half that claimed by that great proponent of post-1870 excess, Ashworth.[28]

What had caused the over-spending in the early period? Irving relied heavily on Samuel Laing's Board of Trade paper of 1844, which showed that British railway costs per route-mile were eight times higher than in the USA, and three to four times higher than in Germany, the higher level of spending adding about £8,000 per route-mile. Approximately two-thirds of this sum was attributable to the failure to restrain engineers and lawyers, which led to a comparatively high standard of construction and a tendency to over-optimism about engineering difficulties. The remainder was due to higher promotional expenses and land acquisition costs. It was thus possible to argue that about a quarter of the £530 million spent on railways to 1870 might have been avoided.[29] His findings confirmed the earlier position taken by Gourvish that high capital costs were introduced at an early stage and with long-term consequences for the railways.[30]

Another major plank of scholarship concerns the relationship between the railways and the state. In this area, we reproduce two pieces which deal with circumstances outside the UK. First, there is R. O. Christensen's two-part evaluation of the Indian Government's role in that country's railway performance in the fifty years to 1920. Second, A. C. Mierzejewski examines the relations between the German state and its railways in the inter-war years. Christensen's aim was to assess the impact of state ownership and control on railway performance, with particular reference to commercial operation.

The first part of Christensen's evaluation dealt with issues of financial efficiency and service standards. Christensen made good use of British government papers from the early 1870s to 1921, which frequently expressed concern about congestion, overcrowding and poor levels of performance. After examining all the familiar pitfalls of assessing efficiency, including comparisons with railways in other countries, he found the 'operating ratio' (working costs as a percentage of gross revenue), that rather crude measure of operating efficiency, to be low and relatively stable. It averaged 50 over the period 1870-89, 49 in 1890-1909 and 52, 1910-21.[31] He then went on to examine operating management, observing that Indian railways were comparatively early enthusiasts of the ton-mile statistic, using it to improve freight train loads from the early

1870s. As a result train-miles run per worker increased by 23 per cent from 1881-85 to 1911-14 and ton-miles run per worker increased by 72 per cent over the same period.[32] What occurred, however, was a disparity between financial results and operating standards. Operating productivity, while improving, remained very low by British standards, but this was not reflected in the financial results, because there was no upward pressure on operating costs, and earnings rose in proportion to the traffic carried. The problem surfaced in the multitude of complaints about railway services, which intensified after 1900. Train speeds were low, provision for third-class passengers was inadequate, and freight was delayed by rolling stock shortages. Traffic growth, particularly in the 1890s, played a part in this, but the main reason lay in the inadequacy of capital equipment.

This element is examined in Part II of Christensen's article. Here low levels of capitalisation were linked to a shortage of investible funds. Although railway management strategies were affected by the complexity of ownership and franchising arrangements, the position of the State was critical. Government policy aimed to maximise railway revenue and to ensure that capital investment did not prejudice income streams. From the 1860s the Government took steps to lower fares and rates and from the 1880s to encourage trunk-line competition, processes which encouraged overcrowding and congestion to the point where the system was on the verge of breakdown by the time of the First World War. Moreover, having taken over responsibility for raising capital on the abandonment of the guarantee scheme in 1870, the Government held down necessary capital investment, and was unwilling to make up for the cautious response of private investors.[33]

Mierzejewski's article, published in 1990, concerned itself with the policy of Germany's national railways, the Deutsche Reichsbahn, established in 1920, and the impact of political change. He found that the foundations for that country's impressive railway modernisation were laid during the period of the Weimar Republic. A modernisation strategy was encouraged by fiscal stringency and competitive pressures, a situation which did not change after the Nazis came to power in 1933. At the time Hitler informed the Director General of the Reichsbahn that railways would be subordinate to road transport. Financial support was halted, and the railways were excluded from the Four-Year Plan of 1936. Consequently, the Reichsbahn financed its modernisation, as it had in the past, from internal resources. It was thus quite wrong of the Nazi regime to claim credit for the successes enjoyed by Germany's railways in the 1930s. Nor was there any evidence to suggest that railway managers were modernising in preparation for war. Their battle was with road competition.[34]

Last, but certainly not least, we include two seminal articles linking transport history with business history and, specifically, with the contribution of that guru of business history, Alfred D. Chandler, Jr. Geoffrey Channon's extended review of Chandler's *The Visible Hand* in 1981 examined the role of railways in both the USA and the UK, while Gregory Thompson's criticism of Chandler for neglecting the realities of railway cost accounting caused considerable interest when published in 1991. Both are key texts. Railways have been part of the mainstream concerns of business history ever since Chandler's article, in the *Business History Review* of 1965, proclaimed them to be 'pioneers in modern corporate management'; elsewhere, they are *'The Nation's First Big Business'*; in the *Visible Hand*, they are 'the first modern business enterprises'.³⁵ Channon, in reviewing what is probably Chandler's best book, not only provided a good summary of the author's general theme; he also identified analytical difficulties involved in assessing the significance of the railways as businesses in both the USA and UK. He rightly placed Chandler in the context of the institutional approach of Weber, Veblen, and Berle and Means, but correctly observed that *The Visible Hand* was a piece of traditional, historical narrative, in which there was no room for the cliometrics of the 'new' economic history.

Channon also asked the reader to take the debate forward by comparing the American experience with that of the British. 'Chandler's typology', Channon noted, 'cannot be readily transposed to Britain'. In the latter case, events moved more swiftly, system-building took place in the 1840s, several decades before it was a feature in the United States, according to Chandler, and the chronology and character of the regulatory regimes were very different in the two countries. In any case the whole timing of industrial and corporate change was different. The railways appeared in Britain *after* the initial phase of industrialisation and urbanisation, but some fifty years *before* the emergence of anything which might remotely be called a 'corporate economy'. In Britain, as in America, the railways were the first big industrial businesses, but Gourvish's work on Mark Huish is cited as a counter to the suggestion that progress in management capability and competency proceeded via an unbroken linear route, which Chandler tends to assume. Channon's critique, then, is important for two reasons. First, it introduced the notion of inter-country comparison to Chandler's work, something which was to blossom in *Scale and Scope*. Second, it introduced a critical edge to the reception of Chandler's ideas, which is fashionable in the 1990s, but most definitely was not in 1981.³⁶

Thompson took issue with Chandler for elevating engineers such as Benjamin Latrobe, Daniel McCallum and Edgar Thomson to significant status in the formative stages of America's corporate history. For Chandler these key executives were at the centre stage not only in developing organisational systems appropriate to the effective management of large, dispersed businesses, but also in introducing accounting and statistical control mechanisms to establish effective management accounting practices. In fact, argued Thompson, the railroads' cost accounting practices left much to be desired. While the array of ton-mile and train-mile statistics indicates sophistication, managers who used them were wrong to assume that the greater part of railroad costs was fixed and did not vary significantly with levels of service provision. This erroneous assumption, claimed Thompson, led railroad managements along a dangerous path of competition by level of service. Thompson developed his argument more fully in a monograph on California's passenger rail services in the twentieth century, and in an article on costing 'ignorance' within the Pennsylvania Railroad, also published in the *Journal of Transport History*.³⁷ He may have taken his argument a little too far, but there is no doubt that it provides a valuable corrective to making easy and sweeping assumptions, as Chandler does, about the significance of the railroads in America's corporate history.

Railways remain a popular source of interest and continue to attract academic analysis, not least in the current age of privatisation. The establishment of a railway studies institute at the National Railway Museum at York is further proof of the interest in, and opportunities for, further research. The articles which appear in this volume deal with major research areas which are of interest not only to scholars, but also to opinion-formers and politicians: the contribution which rail transport makes to economic growth; the investment implications of railway operation; the part played by the railways in the emergence of a corporate economy; and the role of the State in influencing the performance of such industries. This scholarship deserves to be read and will surely encourage others to develop the issues which it raises.

Notes

1 Cf. J. R. Kellett, 'Writing on Victorian railways: an essay in nostalgia', *Victorian Studies*, vol. XIII, no. 1 (1969), pp. 90-6.

2 A second volume, edited by Geoffrey Channon, deals in the main with *micro*-economic aspects of railway history.

3 D. Lardner, *Railway Economy: a Treatise on the New Art of Transport, its Management, Prospects and Relations, Commercial, Financial, and Social* (London, 1850); G. R. Hawke, 'The reputation of Dr. Lardner', in G. R. Hawke, *Railways and Economic Growth in England and Wales, 1840-1870* (Oxford, 1970), pp. 93-9.

4 R. D. Baxter, 'Railway expansion and its results', *Journal of the Statistical Society*, vol. XXIX (1866), reprinted in E. M. Carus-Wilson (ed.), *Essays in Economic History*, vol. III (London, 1962), p. 41.

5 B. R. Mitchell, 'The coming of the railway and United Kingdom economic growth', *Journal of Economic History*, vol. XXIV (1964), pp. 315-36; also reprinted in M. C. Reed (ed.), *Railways in the Victorian Economy: Studies in Finance and Economic Growth* (Newton Abbot, 1969), pp. 13-32; R. W. Fogel, *Railroads and American Economic Growth* (Baltimore, Md, 1964).

6 A. Fishlow, *American Railroads and the Transformation of the Ante-Bellum Economy* (Cambridge, Mass., 1965).

7 Hawke, *Railways and Economic Growth*.

8 P. O'Brien, *The New Economic History of the Railways* (London, 1977).

9 T. R. Gourvish, *Railways and the British Economy 1830-1914* (London, 1980, reprinted 1986, 1989).

10 P. O'Brien (ed.), *Railways and the Economic Development of Western Europe, 1830-1914* (London, 1983).

11 Theme A2b: Inland Transport and Communication from the Eleventh to Twentieth Century: the Industrial Period, August 1990. The participants included Geoffrey Channon, Rainer Fremdling, Antonio Gomez-Mendoza, Terry Gourvish, Patrick O'Brien and Glenn Porter.

12 R. Floud and D. McCloskey (eds), *The Economic History of Britain Since 1700*, 2nd edition, 3 volumes (Cambridge, 1994).

13 This, it should be stressed, is not to impugn his research findings.

14 J. Foreman-Peck, 'Railways and late Victorian economic growth', in J. Foreman-Peck (ed.), *New Perspectives on the Late Victorian Economy: Essays in Quantitative Economic History 1860-1914* (Cambridge, 1990), pp. 73-95; J. S. Foreman-Peck, 'Natural monopoly and railway policy in the nineteenth century', *Oxford Economic Papers*, vol. XXXIX (1987), pp. 699-718; J. Foreman-Peck and R. Millward, *Public and Private Ownership of British Industry* (Oxford, 1994), pp. 81-96.

15 D. H. Aldcroft, 'Railways and economic growth: a review article', *The Journal of Transport History*, 2nd series, vol. I, no. 2 (1972), p. 238 (reprinted in this volume).

16 W. Vamplew, 'Nihilistic impressions of British railway history', in D. N. McCloskey (ed.), *Essays on a Mature Economy: Britain after 1840* (1971), pp. 345-58; W. Vamplew, 'Railways and the transformation of the Scottish Economy', *Economic History Review*, 2nd series, vol. XXIV, no. 1 (1971), p. 38.

17 W. Vamplew, 'Railways and the Scottish transport system in the nineteenth century', *The Journal of Transport History*, 2nd series, vol. I, no. 1 (1972), p. 140 (reprinted in this volume), and see J. Armstrong,'The role of coastal shipping in UK transport: an estimate of comparative traffic movements in 1910', *The Journal of Transport History*, 3rd series, vol. X, no. 2 (1989), p. 164ff., reprinted in the Scolar volume on *Coastal and Short Sea Shipping*.

18 Vamplew, 'Nihilistic impressions', esp. pp. 356-7; 'Railways and the Transformation of the Scottish economy', pp. 37-54.

19 R. W. Fogel, 'Notes on the social saving controversy', *Journal of Economic History*, vol. XXXIX (1979), pp. 1-54.

20 D. L. Lightner, 'Railroads and the American economy: the Fogel thesis in retrospect', *The Journal of Transport History*, 3rd series vol. IV (1983), p. 29 (reprinted in this volume).

21 Ibid., p. 30.

22 J. J. Pincus, 'Railways and land values', *The Journal of Transport History*, 3rd series, vol. IV, no. 2 (1983), p. 36 (reprinted in this volume).

23 Ibid., p. 47.

24 P. Cain, 'Private enterprise or public utility? Output, pricing and investment in English and Welsh railways, 1870-1914', *The Journal of Transport History*, 3rd series, vol. I, no. 2 (1980), p. 16 (reprinted in this volume); Hawke, *Railways and Economic Growth*, p. 92.

25 Foreman-Peck, 'Railways and late Victorian economic growth', pp. 77-81.

26 Cain, 'Private enterprise', p. 23, and cf. W. J. Ashworth, *An Economic History of England 1870-1939* (London, 1960), pp. 120-6; R. J. Irving, 'The profitability and performance of British railways, 1870-1914', *Economic History Review*, 2nd series, vol. XXXI, no. 1 (1978), pp. 46-66; T. R. Gourvish, 'The performance of British railway management after 1860: the railways of Watkin and Forbes', *Business History*, vol. XX, no. 2 (1978), pp. 186-200. It should be noted that the tenor of Foreman-Peck's recent work is that railway efficiency would have been much improved if state planning had been more, not less, vigorous. Cf. Foreman-Peck, 'Natural monopoly', pp. 717-18.

27 R. J. Irving, 'The capitalisation of Britain's railways, 1830-1914', *The Journal of Transport History*, 3rd series, vol. V, no. 2 (1984), pp. 3, 18 (reprinted in this volume).
28 Ibid., pp. 8-9; Ashworth, *Economic History of England*, ch. 5.
29 Irving, 'Capitalisation', pp. 12-18; T. R. Gourvish, 'Railways 1830-70: the formative years', in M. J. Freeman and D. H. Aldcroft (eds), *Transport in Victorian Britain* (Manchester, 1988), pp. 62-3.
30 Irving, 'Capitalisation', p. 7; T. R. Gourvish, *Mark Huish and the London and North Western Railway* (Leicester, 1972), ch. 1.
31 R. O. Christensen, 'The state and indian railway performance, 1870-1920. Part I. Financial efficiency and standards of service', *The Journal of Transport History*, 3rd series, vol. II, no. 2 (1981), pp. 4, 12-13 (reprinted in this volume).
32 Ibid., p. 8.
33 R. O. Christensen, 'The state and Indian railway performance, 1870-1920. Part II. The Government, rating policy and capital funding', *The Journal of Transport History*, 3rd series, vol. III, no. 1 (1982), pp. 25-32 (reprinted in this volume). Under the capital guarantee scheme private capital was attracted by the Government's guarantee of a minimum rate of return of 5 per cent irrespective of a railway's financial performance.
34 A. C. Mierzejewski, 'The Deutsche Reichsbahn and Germany's supply of coal, 1939-45', *The Journal of Transport History*, 3rd series, vol. VIII, no. 2 (1987), pp. 55-6 (reprinted in this volume).
35 A. D. Chandler, Jr, 'The railroads: pioneers in modern corporate management', *Business History Review*, vol. XXXIX, (1965), pp. 16-40; *idem* (ed.), *The Railroads: the Nation's First Big Business* (New York, 1965); *idem*, *The Visible Hand: the Managerial Revolution in American Business* (Cambridge, Mass., 1977), Part II.
36 A. D. Chandler, Jr, *Scale and Scope. The Dynamics of Industrial Capitalism* (Cambridge, Mass., 1990); G. Channon, 'A. D. Chandler's 'visible hand' in transport history', *The Journal of Transport History*, 3rd series, vol. II, no. 1 (1981), pp. 58-62 (reprinted in this volume); Gourvish, *Mark Huish*.
37 G. L. Thompson, 'Myth and rationality in management decision-making: the evolution of American railroad product costing, 1870-1970', *The Journal of Transport History*, 3rd series, vol. XII, no. 1 (1991), pp. 1-10 (reprinted in this volume); idem, 'How cost ignorance derailed the Pennsylvania railroad's efforts to save its passenger service, 1929-61', *The Journal of Transport History*, 3rd series, vol. XVI. 2 (1995), pp. 134-58; *idem, The Passenger Train in the Motor Age: California's Rail and Bus Services, 1910-41* (Columbus, Ohio, 1993).

1 Railways and economic growth[1]: a review article

D. H. ALDCROFT

THE last decade or so has witnessed one of the liveliest debates among economic historians to date, namely the controversy concerning the new economic history or the debate about methodology in history. Indeed almost as much has been written about the virtues or otherwise of the 'new' methodology as on the subject itself. The debate is now becoming somewhat sterile and it is not the purpose here to add to the growing volume of literature in this respect. But for the sake of clarity it is important to get the record straight by way of a few observations about the 'new' art, or should we say science?*

Briefly and simply stated the new economic history lays primary emphasis on measurement and pays specific recognition to the relationships between measurement and theory. It involves the application of economic, statistical and accounting tools to specific problems of economic development. The economic tools provide the means by which to construct a model or theory which can then be tested for its validity by the application of statistical analysis. There is nothing very novel in this procedure as far as contemporary problems or those of the recent past are concerned. To take a simple example: economic theory postulates that there is some positive relationship between production and productivity movements. The validity of this proposition can be, and has been, tested by applying statistical analysis to the observed data.

Clearly the methodological techniques are anything but new.[2] Perhaps the chief departure lies in the fact that in recent years young and energetic scholars, mainly American, have been applying known techniques on an extensive scale to problems of specification in major areas of historical debate. Thus to try and determine the endless and inconclusive descriptive debates about such major issues as the economics of slavery, the role of the railroads in American development, and the impact of declining exports on the growth of the British economy in the later nineteenth century, the new economic historians have sought to formulate explicit propositions and have then proceeded to test their validity. Correct specification is a crucial factor in such studies and for this reason theory is used to identify the explanatory variables for which quantitative information is required and sometimes to derive evidence about those variables for which information is not readily available.[3] Frequently the device of a simulation model is employed to determine what might have happened had certain events not taken place. By constructing a counterfactual situation or hypothetical past, for which invented data are supplied, it is possible to measure the difference between observed data (that is reality) and the predictions of the model.[4] A well-known example has been the attempt to project the likely developments in transportation and economic activity in America in the absence of railroads.[5]

---

* I appreciate the valuable comments made by Professor H. J. Dyos.

That these studies have yielded more than trivial findings is testified by the results so far produced.[6] Other branches of history have not remained unaffected by such developments – they have certainly been caught up in the trend towards quantification – but for the most part the use of advanced techniques has been confined to economic history.[7] In so far as historians have used quantitative techniques at all they have used only descriptive statistics and fairly elementary methods and have tended to eschew problems involving model building and the use of sophisticated statistical techniques.[8] It is evident however that a more scientific approach to the analysis of historical problems is emerging, a trend no doubt encouraged by similar advancements in related disciplines.[9]

Such developments have not gone unchallenged, though given the rather innocuous and yet potentially rewarding nature of the 'new' methodology it is sometimes difficult to comprehend what all the fuss is about. Unfortunately the general debate has centred not on the important issues such as the nature of the methods and tools to be used in such analyses; rather it has either taken the form of a somewhat sterile discussion on peripheral issues, e.g. whether this particular type of history is new or not, or else it has consisted of a frontal attack on the whole concept of the new economic history. The latter has been couched in terms of a basic incompatibility between theory and history but so far it has failed to register any really convincing arguments to justify rejection of the basic methodology. Fritz Redlich, one of the severest critics, argues that econometric history is simply quasi-history on the grounds that it involves the use of hypothetical models or counterfactual propositions which are alien to history, that in turn these cannot be verified precisely, and finally that the methods used are anti-empirical.[10] None of these points of criticism can stand the test of close scrutiny. The substance of the argument rests on the line against counterfactuals but these are germane to all historical enquiry whatever the mode of investigation. Few historians could claim that at one time or another they have not posed the question "What if such and such had not happened then . . . ?" Or to put it in a more concrete form: "What would have happened to the British economy if there had been no return to the gold standard in 1925 (or alternatively, a return at a lower parity)?" On numerous occasions historians have asked this particular question, but the answers have consisted largely of a series of intuitive guesses as to what might have taken place. No specific solution can be derived until the proposition is put to the econometric test.

In other words, to reject the use of the counterfactual would be to destroy much of the base of historical enquiry. The opponents of the new approach are clearly on weak ground therefore in taking stance against the counterfactual concept, especially since, as Murphy has pointed out, "the whole purpose of testing counterfactuals . . . has been to please historians".[11] And after all, the new economic historians have not been engaged in constructing weird and wonderful counterfactuals that are so abstruse as to be divorced completely from reality, but have primarily concerned themselves with making explicit and testing those which have been thrown up by traditional historical enquiry. Thus as Professor Fogel, the leading practitioner of econometric history, makes clear: "the difference between the old and new economic history is not the frequency with which one encounters counterfactual propositions, but the extent to which such propositions are made explicit. The old economic history abounds in disguised counterfactual assertions".[12]

Critics have also argued that quantitative and econometric methods are of limited value because of the inadequacy of the historical data. Such methods are obviously not appropriate to every historical situation and where historical data are inadequate or imprecise their application would yield limited results. Yet while few would contend that a scientific approach can get over this problem satisfactorily "... it hardly follows that, when the sources are suspect or the facts incomplete, an impressionistic, subjective approach can surmount these difficulties".[13] In any case it is perhaps insufficiently appreciated that the use of scientific tools provides scope for a certain amount of data reconstruction and enables the maximum use to be made of what information is available. Thus, if the historical data are satisfactory, then to contend that impressionistic and subjective judgments are better than the findings derived from a scientific analysis of the facts is surely going too far. Those historians who cannot accept this proposition are simply confirming that they do not wish to make the best use of their resources. While at the other extreme, where the data are completely unsuitable for scientific analysis it is questionable whether they can be used very fruitfully by any type of historian.

If the critics of the new approach have failed to come up with any powerful arguments against it this does not mean that the findings and methods of the new economic history have to be accepted without question.[14] In due course some of the results achieved may be subject to modification in the light of new data or the application of more powerful tools of analysis, while the methods themselves are constantly open to refinement. But whatever its present limitations it is clear that the new approach offers the prospect of obtaining far more definitive answers to some of the hitherto unresolved historical problems regardless of whether or not all historians are anxious to know the solutions.[15]

Transport history has gained a place of some prominence in the rise of the new economic history. It is true that the first major formulations were made in the late 1950s when Conrad and Meyer investigated the economics of slavery,[16] but the subject really sprang to life in the mid-1960s with the publication of two profound works on U.S. railroads by Fogel and Fishlow.[17] These set out to determine exactly what contribution the railroads made to economic growth by measuring their social saving, i.e. the difference between the actual cost of shipping goods in a particular year (1890 and 1859 respectively) and the alternative cost of freighting the same commodities in the absence of railroads. The results came as something of a shock to many historians who had previously believed in the overwhelming importance of railways as an agent of economic growth. At the most the social saving derived from railroads amounted to between 4 and 6 per cent of national income, representing approximately one year's delay or more in the growth of the economy.[18]

The railways have been made the subject of the first new economic history written in this country. Following in the footsteps of Fogel and Fishlow, Dr Hawke has produced a work which will no doubt remain a model of its kind. It is certainly a more wide-ranging and intensive study than that of Fogel, though the approach and method of analysis are very similar. As with Fogel and Fishlow, Hawke's main objective is to determine the importance of railways in the economy of England and Wales. The study is centred on the year 1865 and the questions posed are straightforward: "To what extent did the economy depend on railways in 1865?, to what extent could the national income of 1865 have been attained without the innovation of railways?, and what were the social returns to investment in railways?" These questions he approaches through a calculation of the social saving derived from the railways, measured as the difference between the actual cost of transport services in any one year provided by the railways and the alternative cost of the same services in the absence of railways (p. 31). The main social savings calculations for passenger and freight traffic are produced in Chapters II–VII. The remaining chapters investigate the linkages from railways to other sectors of the economy to see if these resulted in any contribution to economic growth which has not been picked up or recorded in the direct social savings calculations. The last chapter summarizes and assesses the findings.

For convenience the main results may be summarized briefly. In aggregate the railways probably contributed about 10 per cent of the national income in 1865: in other words, had they been dispensed with it would have been necessary to compensate for the loss of about one tenth of the national income. The social savings on freight traffic derived from the railways amounted to 4.1 per cent of the national product (most of which came from coal and minerals), while in the case of passengers the railways increased production possibilities by between 2.6 and 7.1 per cent depending on the assumptions made about the degree of personal comfort. Dr Hawke recognizes that the social savings calculations may understate the impact of the railways if they induced in other sectors of the economy changes that favoured economic growth. Thus additional calculations are required to measure any technological external economies that the railways made available to people and firms acting as other than transport users. For instance, the provision of railway services might induce in a particular industry a greater degree of concentration of production resources, or encourage the adoption of new techniques, and these could result in a fall in real production costs which would not be recorded in the social savings calculations. After a fairly intensive survey of the possible indirect repercussions of the railways on other industries, on the labour force and the influence of railway pricing policy and productivity movements etc., he finds that little addition needs to be made for externalities, so that in the absence of the railways the loss to national income would not have been much greater than 10 per cent. Finally, the railways provided a socially rewarding, if financially unattractive, investment; the social internal rate of return averaged 15–20 per cent between 1830–70, a return not likely to be exceeded by an alternative use of resources.

These then are the basic findings. At this point many obvious questions spring to mind: What is their significance? How reliable are the results? Is the method of calculation sound? It would however require more than a short article to do full justice to the volume, for after reading each chapter several queries and criticisms are immediately raised. I shall therefore focus attention on one or two broad issues. First it is essential to assess the significance of the findings and their comparability with American experience. Second, the reliability of the data used and the calculations themselves require some comment. And third, some observations will be made on the basic methodology and assumptions underlying the study.

If we accept the finding that dispensing with the railways would have involved the loss of 10 per cent of the national income in 1865, we must then assess the significance of this figure. Hawke only addresses himself very briefly to this issue and admits, perhaps somewhat grudgingly, that the railways did have a considerable impact on the economy. Unfortunately we have no other yardstick, at least within the context of the British economy, with which to make an assessment, but from our knowledge of the structure of the Victorian economy it seems unlikely that any other innovation or sector could have had such a large impact on the economy. One of the largest sectors in the later nineteenth century was building and construction, but this rarely accounted for more than about 4 per cent of national income and the social return from this sector was hardly likely to have been larger than that of the railways. However, as Hawke notes, this does not necessarily imply that the railways were crucial to the development of a wide range of other industries.

Many writers, when reviewing the work, have naturally enough made direct reference to the U.S. results, and even Hawke himself makes the comparison. Against Fogel's 5 per cent saving for 1890 the 10 per cent for this country appears large. There are of course obvious reasons why there should be a difference: for example, water transport was much cheaper relative to rail in America than in England. Even so it is doubtful whether any really meaningful comparison can be made since the bases on which the two sets of calculations are made are somewhat different. Originally Fogel computed the social saving on agricultural commodities and it was only in a later article that he crudely estimated the savings on non-agricultural commodities and adopted Fishlow's figures for savings in passenger traffic.[19] Thus his aggregate saving of some 5 per cent of the national income cannot be regarded as a very firm result. Moreover, unlike Hawke who does not specify what changes might have taken place in non-rail transportation had the railways not been built, Fogel postulates feasible extensions to the canal network in America. The effects of these are apparently embodied in the final social savings calculations and this would inevitably lead to a lower value compared with that for England and Wales. For these reasons it would be unwise to press the comparison of the results unduly.

In any case, despite Hawke's impressive and wide-ranging inquiry one cannot help but feel somewhat sceptical about the end product. It may be that he has arrived at about the right order of magnitude in his social savings calculations but this could be more through fortuitous accident than conscious design. This is not intended as a criticism of the author's scholarship or the care and lucidity with which he presents his findings. It represents, rather, doubts as to the procedure and assumptions adopted in the study. These can be conveniently dealt with under the following headings: (1) coverage, (2) use of data and (3) methodology.

Although much more comprehensive than Fogel's original study, Dr Hawke's analysis still contains some gaps, the effect of which is probably to underestimate the value of the social savings. For example, no evaluation is made of the productive time savings as a result of faster travel by rail, while the economic effects of carrying mail are not even considered. More important, the study is incomplete in two respects: first, the author does not examine all the leads and lags between railways and other sectors of the economy; and second, no attempt is made to show what the economy would have been like had the railways not been created. Possibly this is asking too much of any one author to undertake but until we have answers on these points a final assessment of the impact of the railways cannot be made.

As far as the first point is concerned Dr Hawke does investigate some of the linkages in the second part of his study: these include the effects of the railways on the coal, iron and a few other industries, and on the capital market, the location of industry and the labour supply. For the most part he finds that the railways were not crucial to the development of other sectors and that most of the gains are already included in the direct social savings calculations. It is doubtful however whether all the technological externalities have been taken into account properly. The brick, stone and timber industries are dismissed in a couple of pages, while engineering and construction are barely considered. Even for those sectors which are discussed in some detail one gets the impression that Dr Hawke is too readily inclined to dismiss their significance. The influence of the railways on managerial techniques and on the development of the capital market and the investing habit, for instance, are probably underestimated.[20] Moreover, although he considers the additional gains over and above the social savings calculation to have been small, no precise estimates are given and the final result simply contains a rather vague 'guesstimate' as to what they might be. Finally, some aspects are not considered at all, such as the wider implications of the railways on the social and urban development of the country. It is possible that the larger town and urban groupings resulting from railway development led to economies in the administration of local governments which are not reflected in the social savings calculations. As Fogel recognizes, "No evaluation of the impact of railroads on [American] development can be complete without a consideration of the cultural, political, military and social consequences of such an innovation".[21]

The above considerations suggest that the social savings may have been underestimated. The implications of the second point noted above are more difficult to determine. Hawke does not hypothesize what would have happened in a non-rail economy, though such an enquiry is crucial to the interpretation of the real social savings derived from the railways. It is probable that in the absence of the latter resources would have been devoted to expanding the non-rail transport network (canals, coastal shipping and roads), and depending on whether this resulted in increasing, falling or constant costs compared with those under a rail/non-rail transport system the social savings would be increased, decreased or remain the same respectively. It is difficult to determine precisely what would have happened since so much depends on the assumptions made. American experience suggests that the social savings would have been reduced given a further extension of the canal network, though this is probably not a very useful guide in view of the wide discrepancy in cost structures between water and rail transport between the two countries.[22] However, it seems plausible to assume that extensions and innovations in the non-rail transport sector (e.g. the development of more efficient road vehicles)[23] would have resulted in falling costs in this sector and thereby offset some of the potential gains from the introduction of railways.

All this is admittedly hypothetical, but not until we know the alternative resource costs under dynamic conditions can we determine the real social savings derived from building the railways. As Hawke rightly points out in his introductory chapter, the concept of social saving is the *ex post* analogue to the *ex ante* concept used in cost-benefit studies and if the latter had been applied at that time to railway investment the alternative use of resources would have had to be taken into account.[24]

Doubts must also be expressed about the selection of data and the basis of some of the calculations. The yearly estimates of social savings which are made for both passenger and freight traffic for the years 1840–70 (to 1890 in the case of freight) are derived from extrapolations on the base year of 1865. Constant real costs are assumed for alternative modes (mainly canals for freight and coach for passengers) so that the multiplicand remains the same for each year. The estimates of the costs of alternative transport modes are derived in a very crude manner indeed. In the case of freight traffic the cost figures of only two canals are considered, the Leeds & Liverpool, and the Kennet & Avon, and these range from 0.12d. to 0.6d. per ton mile. A vague compromise of 0.4d. is then selected which is supposed to represent "as reasonable an estimate as the evidence presently available permits" (p. 84). To this is added the derived cost of shipping by canal (1.5d.), which again is based on fragmentary scraps of data, and an allowance for slightly longer shipments by canal. The final result produces the hypothetical non-rail cost of 2.3d. per ton-mile which when multiplied by the total ton-mileage gives the non-rail cost of goods transport. The social saving is then derived by subtracting the rail cost from the result.

There are several grounds for concern here. The data on non-rail costs are compiled on the basis of very slender evidence and it is extremely doubtful whether they are reliable enough for the purposes to which they are put. Second, it is assumed that the costs of canal transport remain constant through time and for all commodities so that for each year the cost multiplier is the same for all types of traffic. This may not affect the aggregate social savings calculation materially but it could make a considerable difference to the sectoral breakdown. It is difficult to believe, for instance, that it cost about four times as much to carry coal by canal than by rail. Third, only canal transport is considered as an alternative to rail but some freight also went by road and sea where costs varied considerably compared with the canals. In parts of Lincolnshire, for instance, road cartage of agricultural freight was charged at the rate of 6d. per ton-mile as against 1d. per ton-mile by rail in the early 1840s,[25] while in Scotland the median charges per ton-mile for minerals (1860s) were as follows: carts 5.21d., canals 3.86d., east coast sea routes 1.47d., and west coast sea routes 0.66d.[26] These rates do not of course reflect real costs but if similar orders of magnitude prevailed in costs to those reflected in the rates charged by different modes it suggests that some allowance must be made for non-canal transportation when computing the social savings.

Similar criticisms are applicable with respect to the calculations for passenger traffic. Here the alternative mode is coach travel; water transport is dismissed as an unsuitable alternative though in this case the neglect is probably not one of great moment. But again the cost data for coach travel is based on very limited evidence. Two sources only are used, those of Lardner and the Royal Commission on Railways of 1867. The former equated first-class traffic with inside coach travel and second- and third-class with outside coach travel, but the relevant rates are not specified in detail. The Royal Commission equated first-class rail traffic with 'posting' at a cost of 2s. per mile, second-class traffic with inside coach travel and the rest with outside coach travel at 4d. and 2½d. per mile respectively. These values are assumed to reflect real costs – a somewhat dubious assumption – and the resultant calculations produce lower and upper limits of the social savings in personal travel, a large element of which is derived from increased comfort. Apart from the crudeness of the data used it is questionable whether the corresponding classes between rail and coach are correctly specified. Given the fact that all forms of rail travel were superior in speed and comfort to any by road and, in addition, that no allowance is made for productive time savings, then even the upper estimates probably underestimate the social savings as presently calculated. Finally, both in the case of freight and passenger traffic Hawke uses a constant cost money multiplier for conveyance by alternative modes but the railway receipts (costs) are kept in current prices.

Other criticisms could be made on this score, e.g. regarding the derivation of the ton-mileage and passenger-mileage statistics. Much of the data is derived from fragmentary scraps of evidence or extrapolations from single 'authoritative' estimates, and it is doubtful whether they are always adequate for the purposes at hand. The author acknowledges the paucity of information (though in some cases he has not cast his net very wide) and the inexactitude of some of his estimates. Nevertheless, he expresses a surprising degree of confidence in some of the results: "It would be surprising", he says with reference to his estimates for rail output, "if later research shows any of these figures to be greatly misleading" (p. 77).

We have already touched upon some of the more detailed points of analysis and it now remains to discuss the methodological problem in broader terms. In some respects this is the issue which gives grounds for most concern. One reason for this is that Dr Hawke never really explicitly states his methodological approach in detail at the start.

Admittedly he follows close on the heels of Fogel and Fishlow and there is a very useful appraisal of their techniques in the introductory chapter. But given the criticisms which he and other writers have made about their methods of analysis and assumptions it is a pity that he does not clarify the issue for his own subsequent analysis. Instead the reader is left to pick up the assumptions one by one as he works through the chapters and even then they are not always made very explicitly. As a result some methodological ambiguities arise which have quite serious implications.

In order to establish the benefits of an innovation such as the railways it is necessary to know the marginal costs of both rail and all alternative transport modes. It is then possible to derive a value for the real resource saving of the innovation, as opposed to the direct financial savings to transport users, as follows: $(MC_a - MC_r)Q$ where

MC_a = the marginal cost of carrying traffic on alternative modes
MC_r = the marginal cost of carrying traffic by rail
Q = the quantity of traffic to be carried.

However, though relatively simple in theory[27] there are certain practical difficulties involved in the application. Normally, and especially for the nineteenth century, it is difficult to determine the structure of marginal costs in transport very accurately and so it is often necessary to use some price variable as a proxy for marginal costs. This would be all right so long as prices charged by all carriers were equal to their respective marginal costs but this is unlikely to be the case. Second, if an accurate estimate of social savings is to be made some assumption must be made about the likely movement of marginal costs as traffic is diverted from one transport mode to another. For small incremental shifts at the margin this presumably would not pose much of a problem, but for large diversions in any one year it would not be possible to extrapolate from the existing cost data.

It would appear that Hawke makes several assumptions with regard to costs: that the relevant costs used reflect marginal costs of carrying traffic; that rail receipts are equivalent to rail resource costs; and that costs per ton-mile on alternative modes remain constant. Now these assumptions are either questionable or inconsistently applied throughout the study. There is no explicit statement on the marginal cost concept and the treatment of this problem varies from one transport mode to another. In the case of canal transport some attempt is made to distinguish between non-rail charges and non-rail costs though whether the latter can be regarded as the true marginal cost for canals, given the crude manner in which it is estimated, is another matter. However, for road and rail transport marginal costs are assumed to be equivalent to rates charged. Now although for road transport this may have been true in some cases, at least where rail competition forced road charges down towards marginal costs, for the railways this was clearly not so unless they were making zero profits. In any case, since the railways did not know the marginal costs of their individual services "it is still less plausible to relate the average of the individual rates to marginal costs" (p 291); in which case it is surely unrealistic to use rail receipts as a measure of resource costs. The final assumption, that of constant costs of non-rail transport, is particularly dubious. Given the increasing diversion of traffic to alternative modes it is clearly somewhat implausible to assume that their costs remained constant, but this brings us back again to a consideration of the changes which would have occurred in the non-rail sector in the absence of railways.

There can be no denying that Dr Hawke has produced a most scholarly and intellectually stimulating volume which will provide a focus of debate for some years to come. Furthermore, it is also a very informative piece of work containing much of interest to those readers who are not specially interested in quantitative and econometric techniques of analysis. There is a surprising amount of information and factual material (some of it not always very relevant to the main theme of the volume) on all manner of topics, including not only the railways themselves, but also employment, agriculture, various industries – especially iron – and the capital market. In fact the book would provide a very useful starting point for a study of the mid-Victorian economy, though its value in this respect is certainly not enhanced by the derisory index of less than three pages.

Nevertheless, in the light of the above considerations it can be argued that a precise estimate of the social savings from the railways still remains to be made. The study leaves several points open to question which give rise to some doubt about the credibility of the findings. Without performing the necessary calculations it is impossible to say whether Dr Hawke has overestimated or understated the gains from the railways, but it is clear that considerable refinement and modification of the analysis are required before we can be confident that we have the final result. It may be, as one critic of Fogel has hinted, that it is asking too much to ask "What if there had been no railroads ...?"[28] Whether this is correct or not one thing is certain: the railways will never be the same again since the advent of Fogel, Fishlow and Hawke.

NOTES

1. G. R. Hawke, *Railways and Economic Growth in England and Wales, 1840-1870* (Clarendon Press: Oxford, 1970. xiv + 417 pp. £6·00).
2. Even with respect to historical problems. Before the war Beveridge, for instance, was using regression analysis in his investigations of the trade cycle.
3. For a lucid exposition of the techniques see H. J. Habakkuk, 'Economic History and Economic Theory', *Daedalus*, c (1971).
4. The 'newness' of this is not the use of the counterfactual, which is implicitly incorporated in much traditional historical enquiry, but the fact that the counterfactual is made explicit and quantified. M. Desai, 'Some Issues in Econometric History', *Economic History Review*, xxi (1968).
5. R. W. Fogel, *Railroads and American Economic Growth: Essays in Econometric History* (Baltimore, 1964). So far most of the counterfactual exercises relate to America though Smelser attempted a reconstruction of the social structure of Lancashire in the absence of an industrial revolution. N. J. Smelser, *Social Change in the Industrial Revolution: An Application of Theory to the Lancashire Cotton Industry, 1770-1840* (1959).

6. The literature is too extensive to quote in full but a useful collective survey of recent work can be found in R. W. Fogel and S. L. Engerman, *The Reinterpretation of American Economic History* (New York, 1971). See also the papers and bibliographical material in R. L. Andreano (ed.), *The New Economic History* (New York, 1970). For a short and useful summary of current achievements and the scope for further enquiry see A. Fishlow and R. W. Fogel, 'Quantitative Economic History: An Interim Evaluation: Past Trends and Present Tendencies', *Journal of Economic History*, XXXI (March 1971).

7. J. M. Price, 'Recent Quantitative Work in History: A Survey of the Main Trends', *History and Theory*, Beiheft 9 (1969). Quantitative work is nothing new in economic history and some economists have turned their hand to this kind of work, though many economic historians prefer to regard this as a branch of applied economics.

8. W. O. Aydelotte, *Quantification in History* (Reading, Mass., 1971), 28. Though Professor Aydelotte's conception of what is involved is perhaps somewhat oversimplified (see p. 44).

9. In urban history, for instance, more explicit questions are being asked and new techniques adopted including the application of sociological theory and the use of quantitative materials and statistical methods. See S. Thernstrom and R. Sennett, *Nineteenth-Century Cities: Essays in the New Urban History* (Yale, 1969), vii, and S. Thernstrom, 'Reflections on the New Urban History', *Daedalus*, C (1971), 370. In other disciplines, notably geography, the use of more scientific techniques of analysis has made considerable progress. See H. C. Prince, 'Real, Imagined and Abstract Worlds of the Past', *Progress in Geography*, III (1971), for an extensive review of recent research.

10. F. Redlich, ' "New" and Traditional Approaches to Economic History and their Interdependence', *Journal of Economic History*, XXV (1965).

11. G. G. S. Murphy, 'On Counterfactual Propositions', *History and Theory*, Beiheft 9 (1969), 16.

12. R. W. Fogel, 'The New Economic History: Its Findings and Methods', *Economic History Review*, XIX (1966), 655.

13. Aydelotte, *op. cit.*, 52.

14. Nor should the above argument be construed as an attempt to reject the utility of traditional economic history. Quantitative analysis clearly has limits as well as limitations and it is certainly unlikely to produce much technological obsolescence among traditional economic historians. In any case the polarization of the two groups has surely been pressed too far, and both have much to offer each other. An impassioned plea for more co-operative effort and the termination of the divide between the two opposing methodological camps has recently been made by H. W. Richardson, in 'British Emigration and Overseas Investment, 1870–1914', *Economic History Review*, XXV (1972), 100, 108.

15. It would seem that some are not. Professor A. Schlesinger, Jnr, for instance, has written that "almost all important questions are important precisely because they are not susceptible to quantitative answers". – 'The Humanist Looks at Empirical Social Research', *American Sociological Review*, XXVII (1962), 770.

16. A. H. Conrad and J. R. Meyer, 'The Economics of Slavery in the Ante-Bellum South', *Journal of Political Economy*, 66 (1958) and 'Economic Theory, Statistical Inference and Economic History', *Journal of Economic History*, 17 (1957).

17. Fogel, *op. cit.*; A. Fishlow, *American Railroads and the Transformation of the Ante-Bellum Economy* (Cambridge, Mass., 1965).

18. This represents a very emasculated summary of the main findings, some of the points of which will be elaborated below.

19. R. W. Fogel, 'Railroads as an Analogy to the Space Effort: Some Economic Aspects', *Economic Journal*, LXXVI (1966), 39–40.

20. See T. R. Gourvish, 'Captain Mark Huish: A Pioneer in the Development of Railway Management', *Business History*, XII (1970), and J. R. Killick and W. A. Thomas, 'The Provincial Stock Exchanges, 1830–1870', *Economic History Review*, XXIII (1970).

21. Fogel, *Economic Journal*, loc. cit., 40 n. 2.

22. In the United States water and rail costs for freight were very similar whereas this was not the case in England and Wales.

23. This is not so impractical as it sounds. By the early 1830s steam carriages had been developed to the extent that they offered one of the cheapest and quickest forms of travel over short distances. Had the railways not appeared it is conceivable that much greater improvement in the mode of land carriage would have taken place. See Walter Hancock, *Narrative of Twelve Years' Experiments (1824–1836) Demonstrative of the Practicability and Advantages of Employing Steam Carriages on Common Roads* (1838), 3–5, 97.

24. On this point see Paul A. David, 'Transport Innovation and Economic Growth: Professor Fogel on and off the Rails', *Economic History Review*, XXII (1969), 513.

25. Samuel Sidney, *Railways and Agriculture in North Lincolnshire. Rough Notes of a Ride over the Track of the Manchester, Sheffield, Lincolnshire and Other Railways* (1848), 13.

26. Wray Vamplew, 'Railways and the Transformation of the Scottish Economy', *Economic History Review*, XXIV (1971), 51.

27. To obtain a precise formulation for social savings certain other variables need to be taken into account but for present purposes these need not concern us here. See P. D. McClelland, 'Railroads, American Growth, and the New Economic History: A Critique', *Journal of Economic History*, XXVIII (1968).

28. McClelland, *loc. cit.*, 121.

2 Railways and the Scottish transport system in the nineteenth century

W. VAMPLEW

I

"Canals and turnpike roads have, as it were, had their day, and must yield the palm to that greatest achievement of modern times, the iron railway."[1] This verdict of one contemporary of the early railway age was clearly accepted by the transport historian, W. T. Jackman: "Within the first twenty years of the railway era, this young giant had overshadowed all other systems of carrying, some of which had taken centuries for development."[2] This article examines the validity of these opinions as applied to Scottish railways: first by a brief evaluation of the advantages that railways held over other modes of transport, and then by assessing the effects in aggregate of these advantages on roads and road-users, canals, and coastal and river shipping.

Railways were not always directly cheaper than alternative forms of transport, especially those concerned with water. Railway promoters, however, argued that railways could bring substantial savings because of their greater speed, regularity and convenience. Without doubt railways were the fastest means of transport to be widely adopted in the nineteenth century. This had obvious advantages for passenger travel: Sidney Smith's "early Scotsman" may have had to pay extra to "scratch himself in the morning mists of the north and have his porridge in Piccadilly before the setting sun", but, as a correspondent to the *Railway Times* pointed out, "the Scots are no doubt strict economists, but they understand the economy of time as well as that of money".[3] Table 1 suggests that on at least two routes in the early 'forties Scots exhibited a preference for faster if more expensive travel. However, in the absence of data for other years and without knowledge of relative capacity utilization, comfort or safety, we cannot be too definite on this issue. The speed of the railways also benefited freight traffic: perishable commodities became more marketable and,

TABLE I

Route	Transport	Number of Passengers	Fare	Time
Glasgow–Paisley	Canal	185,000	4½d.	60 mins
	Railway	423,000	8d.	15 mins
Glasgow–Greenock	Canal	312,000	8d.	150 mins
	Railway	611,000	12d.	60 mins

Source: S.C. Railways; 1844 XI.

where suppliers could respond readily to demand changes, less stocks had to be held.

Regularity rather than speed appealed to others. Charles MacLaren, one-time editor of the *Scotsman* and a driving force in getting railways accepted in Scotland, maintained that railways were less affected than canals by frost or drought and that, unlike coastal shipping, commodities sent by rail would not be "detained weeks on end by wind and tide".[4] As to cart transport, Henry Brown, a woollen manufacturer of Galashiels, writing in 1829, advised the laying in of coal stocks for the winter because a combination of bad weather and worse roads made transport difficult at that time of the year.[5] Yet were the railways so exempt from the vagaries of the Scottish climate as their supporters believed? Embankments washed away by heavy rains, rails buckled by the sun and bridges damaged by ice floes may have been infrequent occurrences, but not so snow drifts. Every winter saw some part of the Highland system immobilized. On one famous occasion in 1879 it was announced by officials in Inverness on the Monday night that the whereabouts of the 3.10 p.m. Sunday Mail was "an absolute mystery".[6] Yet it is arguable that these climatic conditions would equally have stopped other forms of transport, for during several extreme winters the railway journals made great play with the fact that the railways got through when all else failed.

Allied to the question of regularity is that of convenience. Here MacLaren made two important points: first, "railways have this grand advantage, that they may either be usefully combined with the ordinary roads of a country, or be substituted for them, and expanded into a general system of internal communication"; and, second, farmers would not cart to canals "with the trouble of attending shipping and unshipping".[7] Not only farmers gained from the flexibility of the railways: industrialists found that a railway line could easily be laid into their factory yards to run where it was required and thus save the costs of double loading and unloading which a combination of cart and canal would require. Even where journeys involved both railways and carts, the flexibility of the locomotive track reduced the volume of relatively expensive horse haulage that would have been required had the journey been cart alone or some combination of cart and waterway.[8] Canals frequently proved incapable of adaptation to new conditions with the result that unenlarged locks and aqueducts prevented modern sea-going vessels from using the canals and thus involved the coastal steamers in additional transhipment of goods.[9] In turn coastal steam shipping may have offered more regular service than sailing vessels but still many coastal routes were sailed only once or twice a week. Admittedly this was a consequence of low demand for the service, but it did mean that a long delay was incurred if a load or passenger missed the boat. Railways, partly because of the intense inter-company rivalry, offered a much more frequent service.[10]

II

When a railway challenged a turnpike there was no doubting the victor. It is clear from Table 2 that the construction of competitive railways adversely affected the financial position of turnpikes. Between 1834-5 and 1848-9 trusts competing with railways experienced a fall in income more than twice as large as that for Scottish turnpikes as a whole, and in the 1850s, as railways consolidated their networks and built up their traffic, competing turnpikes continued to suffer a greater decline in revenue than turnpikes in general.[11] The traffic estimates submitted by the railway companies in their parliamentary petitions suggest that they were quite confident of attracting most of the traffic travelling on roads parallel to the line. The turnpike trustees too were well aware that the railways were dangerous rivals, and forced many of the railways not only to buy off their opposition but also, in the majority of cases prior to the railway mania, to undertake responsibility for relieving the trustees of a portion of the debt for which they expected to become liable.[12] Yet not all turnpikes were damaged by the advent of a railway in their locality. A Select Committee of 1839 declared that "it appears that nearly all roads or highways leading to stations, or termini, of steam communication have increased in their traffic."[13] There is no reason to assume that this conclusion was invalid for Scotland, especially when there is evidence that a turnpike built to link the Arbroath and Forfar line with Brechin enjoyed a considerable volume of traffic.[14] However, whether the conclusions of 1839 would hold in 1858 when branch lines had weakened the utility of turnpikes as feeders is debatable. Turnpike revenues as a whole were on the decline and it may well be that carrying traffic to a railway station merely retarded the fall for individual roads.

As for the road users, it seems that the railways soon replaced the horse or the horse and gig as a mode of transport for commercial travellers, although contemporary opinion rather than statistics have to be relied upon in maintaining this.[15] As regards carriers and carters, however, some quantifiable information is obtainable from the Edinburgh and Leith Post Office Directories. In 1838/9 the number of journeys made by carriers each week from the metropolis to other towns and cities in Scotland, as well as to several important north of England cities, was 660. This involves some

TABLE 2

Revenue of Scottish Turnpike Trusts

Trusts	1834–5 £	1848–9 £	1858–9 £	% change		
				1834–5 to 1848–9	1834–5 to 1858–9	1848–9 to 1858–9
(a)	254,679	237,696	204,677	6·7	19·6	13·9
(b)	25,612	21,456	13,743	16·1	46·3	35·4
(c)		25,388	17,414			31·4

Source: Parliamentary Papers.
Notes: (a) All Scottish turnpikes.
(b) 13 trusts known to be in competition with railways before 1845.
(c) 18 trusts known to be in competition with railways before 1855.

double counting as a carter might visit more than one town on a trip, but even allowing for this a fall in the number to 231 within 20 years is significant evidence that the railways had a detrimental effect on the long-distance carrier. On the other hand the substantial evidence of contracts made by the railway companies with local carriers coupled with the development of the parcel post suggests a different story where intra-urban cartage is concerned. C. Robb and Company, for example, who had previously acted as general carriers between Glasgow and Kilmarnock, found it more profitable to become carters for the railway companies in Glasgow and other large towns.[16]

. Carriers were of course concerned with goods; passengers were primarily the province of the stage coaches in the pre-railway era. The greater speed of the railway attracted many of the old coach passengers to the new form of travelling and most competitive coaches disappeared virtually overnight, or, in the case of coaches carrying mails, at the end of their post office contract. Such was the overwhelming competitive advantage of the railway that when, in spite of available railway accommodation, the Glasgow to Stirling coach was found to be in high demand, one railway journal remarked that there must be something very wrong with the management of the railway company.[17] Several other coach proprietors realized that they could survive as viable enterprises, at least until the railway network of Scotland was fully mapped out, by running coaches in connection with railway services. Running a more flexible service was also a way of surviving. The Edinburgh to Glasgow coaches, for example, outpaced by the railway, switched to a longer route and tapped areas previously left unserved.[18] In addition, a few coaches were able to take advantage of the unintegrated state of the Scottish railway companies and exploit through routes which defective or non-existent inter-company arrangements left for them. Presumably these opportunities declined as amalgamations and working agreements increased.

Finally it might be noted that the suburban services of the railways which replaced local coach services in the 1870s were themselves the victims of the march of progress towards the end of the century when municipal and private tramways, with their higher route density and more frequent stops, began to compete for the intra-urban passenger traffic.

III

At the end of the nineteenth century Scotland possessed only five canals of any significance, three of which were in the hands of the railway companies. Both the independent canals, the Crinan, built to save vessels a 70-mile voyage round the peninsula of Kintyre on their way to the Clyde from the north-west of that river, and the Caledonian, projected chiefly with a view to facilitating trade between the Baltic and the west coast of Scotland, had to resort to the public purse for completion of their construction. Their economic existence proved as difficult as their birth, for what little potential traffic there might have been was first deterred by the rates charged, and later by the failure of the operators to adapt to the needs of improved and larger-capacity vessels. From the first scarcely viable, and later commercially obsolete, these two canals remained relatively devoid of traffic which might have gone by rail, and thus avoided railway competition or take-over.

The other three major canals, the Monkland, the Forth & Clyde, and the Edinburgh & Glasgow Union, formed a useful coast to coast route in the important industrial belt of the Central Lowlands. These were in the hands of the railway companies, but they had not been taken over to be utilized as a through system for rival railways controlled the canals involved.

The Monkland in the Lanarkshire coalfields possessed a monopoly of the coal trade from that area into Glasgow which it exploited to such an extent as to stimulate proposals in the 1820s for railways designed to break the stranglehold. The canal proprietors fought back, first by opposing the building of the lines, and when this failed by cutting rates, increasing facilities and building short feeder railways, this time meeting with more success. In 1846 this canal was transferred to the Forth & Clyde Navigation, which, as the two were practically one system, seemed a logical step, especially in view of the expected increase in railway competition following the railway mania.

The Forth & Clyde Navigation, linking Grangemouth on the east coast with Bowling on the Clyde, appears to have been a highly successful venture, though once again complaints were voiced about the exploitation of its monopoly position. An additional stimulus to traffic was given in 1822 by the opening of the Union Canal which linked the Forth & Clyde Navigation with the city of Edinburgh. Twenty years later, however, the construction of the Edinburgh & Glasgow Railway took away practically all the passenger traffic and led to a rate war for the goods traffic in which both sides lost heavily. At one stage amalgamation of all the canal concerns with this railway was contemplated, but the opposition of the English shareholders in the railway company to the amount of compensation demanded by the canals, and the opposition of many trading interests to the takeover as such, eventually produced a situation where the Union Canal was absorbed by the railway and an agreement reached with the Forth & Clyde Navigation as to fixing rates and apportioning traffic.

Not until two decades later was the question of takeover again mooted. This time the railway concerned was the Caledonian, which was seeking an east-coast port which it could develop in order to take traffic away from its major rival, the North British, which already had such an outlet at Leith. Grangemouth seemed the obvious choice, but the Caledonian found that it was unable to obtain control of the port without also taking over the Forth & Clyde Navigation. Several factors, however, were working to encourage the Caledonian to do this. First, the canal to some extent offered a challenge to the Edinburgh & Glasgow Railway, a company which had previously been involved in a joint purse agreement with the Caledonian before opting out and amalgamating with the North British. Secondly, although the Caledonian and the Navigation company had come to an arrangement in 1857 "with the view of preventing injurious competition", in 1865 the latter organization was proposing the construction of two new canals which would be in competition with branches of the Caledonian. Finally, the canal was a going concern and the railway directors believed it would pay for itself in the future.[19]

Their expectations were fulfilled in the sense that the canal was not an unprofitable burden, but whether money spent on the canal would have been more profitably employed elsewhere is open to question in view of the fact that the acquisition of the canal by the railway coincided with an accelerated decline in the canal receipts. This does not appear to have been the result of deliberate policy by the railway company, for the Caledonian had every incentive to promote traffic development on the Monkland Canal, situated as it was in a coalfield dominated by the North British and, as stated, the Forth & Clyde also challenged this rival railway. There is no evidence that the railway allowed the canals to stagnate; in fact everything points to a greater degree of efficient maintenance.[20] The decline in traffic was only to be expected as local collieries became exhausted, and as rival railways built branch lines and intensified their efforts to tap off canal traffic. And of course Caledonian branches built in retaliation had the same result so far as the canals were concerned.

The Edinburgh & Glasgow Union not only provided the Scottish metropolis with western coal via the Forth & Clyde Navigation but rivalled the Edinburgh to Glasgow coaches for passenger traffic. With the opening of the railway linking these two cities, however, came a decline from which the canal never recovered. Both the Forth & Clyde and the Union proprietors had strongly opposed the building of the railway and had forced the railway to adopt a steep incline tunnel at the Glasgow end on the grounds that a high-level station might interfere with future plans for a ship canal. Opposition was still strong when the railway came into operation and a period of intense competition ensued in which passenger fares fell by a third and some goods rates by seven-eighths.[21] From the time of the opening of the railway to the absorption of the canal by the railway in 1849, the canal's revenues just covered maintenance costs and interest owing on debts. As far as can be ascertained the railway decided to take over the canal simply to end the ruinous competition with no strong views as to developing the traffic. However, according to dues paid to the City of Edinburgh the traffic was fairly well maintained until the latter decades of the century when it fell away, even though the canal had been kept in good condition and the railway company had made several attempts to stimulate custom by offering low rates.[22]

Despite their declining profitability, it would seem that railway-owned canals in Scotland did well as compared with most canals in the United Kingdom.[23] But the story is incomplete, for Scotland had possessed three other incorporated canals in whose disappearance the railways, as owners and as competitors, played an important role.[24]

In 1807 a canal was begun to link Glasgow with Paisley and Ardrossan, but fund-raising difficulties resulted in its stretching only as far as Johnstone. When the idea was revived in the late 1820s technology had intervened and the connecting link between Ardrossan and Johnstone was to be a railway. This too proved unattractive to investors and reached only as far as Kilwinning. Unfortunately, the construction had been started from Ardrossan which left a situation where the scheme had produced a canal at one end, a railway at the other, and nothing in between to link the two. The inconvenience of having the canal and railway under the same management led to a separation of controlling powers in 1840. This did not make the canal any more profitable, especially as railway competition intensified. The competing railway accepted an agreement as to the division of traffic, but this was ruled illegal by the Law Officers of Scotland, and the railway directors also realized that they could do better without the arrangement. Although earlier they had tried to smash the canal by rate warfare, in 1869 the Glasgow & South Western Railway took over the canal, now heavily in debt, to prevent it falling into Caledonian hands, and were put under an obligation to keep it in good order. Although this was done, the canal facilities were not in demand and the railway decided to use the land occupied by the canal to more economic purpose and obtained permission to use the canal bed for a railway, the so-called 'Paisley Canal Line', which linked up with the rest of their system.[25]

Another canal that came into railway hands and was eventually closed was the ¾-mile-long Forth & Cart, which, by acting as a link, gave continuous navigation from the Forth & Clyde to the town of Paisley. This canal was taken over by the Forth & Clyde Navigation and thus later came to be a possession of the Caledonian Railway Company. Unfortunately for the Forth & Cart the opening in 1882 of the Glasgow, Yoker & Clydebank Railway under the auspices of the North British took away nearly all its traffic. It struggled on until 1893 when it was decided to close down and lay part of the Lanarkshire & Dunbartonshire Railway along the canal bed.

The Inverurie to Aberdeen canal was the third waterway to form the base for a railway line. This was not a financially successful canal and when in the 1840s the projected Great North of Scotland Railway offered the proprietors £36,000 for it, under the threat of running a railway line parallel to the canal, the offer was taken up. The railway company had a threefold objective in buying the canal. They believed it would save on land costs if their line was laid along the canal bed; they also felt that their line might not have been sanctioned had the purchase not been proposed; and of course filling in the canal removed potential competition.[26] So keen were they to begin construction that the canal was drained before it was clear of vessels!

IV

In contrast to turnpikes and canals, coastal shipping remained a formidable competitor of the railways throughout the nineteenth century. Not until the last decade of the century did railways, by use of exceptional rates, begin to seriously undermine the major traffic of the coasters. To some extent the competitive success of the coasters lay in offering cheap fares on the longer passenger journeys, but primarily it was in goods traffic that they challenged the railways.

That coastal shipping more than held its own (except in perishables and cattle) is clear from statements made by the railway companies and contemporary observers. The Caledonian found in 1850 that "the competition by sea has made it necessary to carry at less remunerative rates than might otherwise be charged". About the same time, but on the other coast, the North British was finding its earlier prophecies that "steamers might pay to the north of Edinburgh, but to the south their attraction could never compete with the rail", had been too optimistic. Later in 1875 the North British declared that it was "quite clear" that in certain classes of goods it could not compete with the steamers. Further north the Highland acted cautiously about any rate increase because of the active competition of sailing vessels.[27] Numerous other examples could be given, all of which would show that the railways were well aware of the realities of competition from the slower but cheaper coasting vessels. Other factors besides cheapness encouraged the use of shipping in preference to the railways. It was argued that in Aberdeen the steamers were more convenient as the railway station was too far out of town. It was also maintained that several of the east coast railway companies were too concerned with taking traffic to the south via the west coast route, presumably because of agreements made with the Caledonian; this, it was said, gave the east coast steamers an advantage. In addition, to send goods by rail down the east coast involved making arrangements with the North British and, at the time (1853), this company was far from being an ideal commercial partner. Ironically, when an Anglo-Scottish railway traffic agreement was organized between major companies on both sides of the border, rates tended to rise and trade was diverted from the railways to the steamers.[28] Free access to the sea meant that individual industrial companies could, if they wanted, own their own vessels. One notable example of this was the famous Carron iron company which set up its own line of coasting screw steamers to London, not only for its own purposes, but also for the carriage of the public's goods, a venture which at times proved quite profitable.[29]

Well aware of the competitive prowess of the coastal steamers, the railway companies made several attempts to come to terms with them. The first attempts were to employ steamers to fill in routes where the railway did not operate. Directors and shareholders of the Glasgow, Paisley, Kilmarnock & Ayr assisted a company setting up a steamboat service between Liverpool and Ardrossan in 1840 and, although this agreement broke down, fresh arrangements seemed to hold, at least until the west coast railway route was operative.

Later arrangements, or attempted arrangements, were of two types. First, a railway company that was challenged directly by coastal shipping would often try to get agreement on the rates to be charged so as to avoid excessive competition. Secondly, some railway companies would make arrangements to convey traffic to the boats sailing in competition with rival railway companies. Neither proved very successful or permanent. The first type broke down because the shipping proprietors felt little need for them except in the northern counties where there is evidence that the coming of the railway eroded the prosperity of the coastal trade.[30] The second kind declined as the smaller railway companies were absorbed into larger concerns and also as through railway arrangements became more common. When railway companies entered the shipping trade it was generally concerned with river traffic but there were some ventures with sea-going vessels. The North British attempted to organize steamers from Silloth to Carlisle, the Portpatrick railway became involved in the Irish Sea ferries, and, in the north, the Highland Railway purchased steamers to establish a service between the Western Isles, Orkney and the mainland. It might be noted that in the latter case the service was taken over within five years by MacBrayne of Glasgow: this was to the railway's satisfaction as they had always opined that their function was to manage railways "and leave the steamers to others", but they could not get anyone interested enough to start the service in the first place.[31]

The river steamers, especially those on the Clyde, also proved formidable competitors to the railways whose solution was to attempt to make agreements and eventually to go into the steamer business themselves. Passengers, especially those on excursions, seem to have preferred cheapness and comfort to speed and reliability. Many of the early agreements were to employ steam boats to run in conjunction with railway services: because the railways were unable to travel the whole distance from Glasgow to various ports served by steamers, they hoped that the superior speed of the railways for part of the journey, coupled with a steamboat for the rest, would enable them to compete successfully against steamers going the full distance. These agreements never seemed to last: it was financially less rewarding for the steamers to travel (say) half the distance and hand over its passengers to the railways than to take them the full journey, especially when, without their co-operation, the railway could not guarantee passengers the full trip.

The answer, so far as the railway companies were concerned, was to purchase or charter vessels and run steamer services themselves. The first to attempt this was the Glasgow, Paisley & Greenock in the mid-1840s, but it ran into tremendous opposition from the existing steamboat owners, as did all the later attempts by other railway companies.[32] Time and time again when the railways attempted to obtain powers to run steamers they were defeated in Parliament.[33] It was argued by the steamboat owners that the railways, in having limited liability, possessed an advantage over the steamboat companies which would allow them to amass capital, compete on unequal terms and, in driving out the independent steamers, obtain a monopoly of the river traffic. Furthermore, it was contended that the railways could make up on their railway charges what they might lose in undercutting the independent boats on the river. However, in 1864 a Select Committee on Railway Companies' Powers was "unable to see why companies owning railways should not own steamers also, if it be expedient for the development of their traffic, why to the business of carrying by land, they should not add that of carrying by sea."[34]

The recommendation of the 1864 Committee merely gave recognition to the realities of the situation, as on several occasions when parliamentary permission to own steamers had been refused, the directors of a railway had agreed to run boats in their name providing that the railway shareholders would guarantee them against loss. Usually the shareholders were willing to give permission; trouble only arose when the steamers were put into operation and the shareholders began to suspect that losses were being incurred which undermined railway dividends. How valid their protests were is difficult to ascertain as the railways did not publish separate accounts for their ancillary concerns; but the very fact that they did not, coupled with the evasive replies often given by directors to shareholders' questions, points to some justification for the complaints. The Committee of Investigation into the affairs of the North British Railway in 1866 admitted that the Steam Packet Company with which the railway was associated had been "an unfortunate and unremunerative undertaking". However, they did add that "whether it may not have indirectly have benefited the railway [they] have not had the means of ascertaining."[35] This is the key point: did the traffic that the steamers brought to the railways more than compensate for the losses on the traffic that the steamers themselves carried? If the railway directors were rational businessmen – though in view of some of their other decisions on capital expenditure this is debatable – then some light on this point might be shed from the histories of the various railway steamer enterprises. If they were allowed to survive then possibly they were making a contribution to the overall profits of the railway company.

The North British Steam Packet Company was set up as a nominally independent concern. This had come about when the railway company had leased the Port Carlisle

Railway, the Silloth Railway and Silloth harbour in 1861, and had, at the same time, purchased the vessels of the Silloth Bay Navigation Company in order to make use of what they had earlier termed "the great facilities for the shipment of coals, minerals and passengers to Ireland and the west ports of England". However, after John Burns, a steamboat owner, obtained an interdict against the company, Parliament gave them permission for railway steamers to ply only between Belfast and Silloth. This led to several of the railway directors making the offer to run a packet company on the guarantee that personal losses would be made up from the railway funds. The packet company continued to operate through the century as a private concern before being officially acquired by the railway in 1902 and was not wound up until after the First World War. Calculations made by the railway company in 1908 suggested that if the steamers were given up over £4,000 would be lost.[36] What this was as a return on capital, however, was not calculated. The North British also had a share with the Caledonian and the Lanarkshire & Dumbarton railways in the Loch Lomond steamers and in addition took over the Galloway Steam Packet Company which was hitting the railway with its Firth of Forth pleasure cruises.

The Caledonian Railway, having absorbed the Glasgow, Paisley & Greenock which had previously tried to break into the steamer business, also made an attempt in 1852, but failed because of the competition of the existing vessels. Their failure in fact led to an agreement with the existing steamboat owners which lasted some 15 years. Not until the late 1880s did the Caledonian, after several parliamentary defeats, decide to adopt the expedient of an 'independent' Caledonian Steam Packet Company. This attempt to break into the river trade seems to have succeeded. It was perhaps because of the racing prowess of Captain Williamson, the Caledonian's man in charge, but more likely it was because economic conditions were propitious with a larger, more travel-minded population in the areas served.[37]

There was less opposition from the steamboat owners to the railways running ferries, partly because as early as 1848 Parliament had laid down the general principle that a sea passage of the nature of a ferry was more convenient to passengers if in the hands of the connecting railways. From then on there was explicit recognition of "the great facilities given to the public by placing all the links of through communications in the same hands." The reservations of the steamer proprietors were confined to comments such as that made by a witness to the 1864 Committee, "from the experience which I have had in Committees, the definition of ferry obtains a very wide range."[38] In general they had little objection to the railways running genuine ferry services and those across the Tay and Forth were in operation as soon as the adjoining railways were opened.

V

Scottish railway companies in the nineteenth century did not confine their activities to their own lines. At various times they involved themselves, either by ownership or by making agreements with existing participants, in almost every other form of transport. To some extent they were pursuing a deliberate policy of establishing co-ordinated transport services, but a more prevalent motive was competition, either that intense rivalry which existed among Scottish railway companies or the struggle, no less fierce, between the railways and other transport media. In the latter competitive battle the railways found that they were not able to sweep everything before them as their promoters had prophesied. By offering cheaper rates than the railways, coastal steamers retained much of their traffic, especially freight, till the end of the century. River steamers, too, were formidable competitors; again it was cheapness that proved decisive, though in this case it was passengers that were attracted. However, the speed, convenience and dependability of the railways so reduced total transport costs that no rates could save long distance carters, stage-coaches, turnpike trusts or canal companies. Some canals facing railway competition did make reasonable profits for quite some time, but this was because their ownership was divided between two rival railway companies, each of which found the other's lines competing against their stretch of canal. This gave them an incentive to encourage the canal traffic. Over time, however, even these canals nourished by the railway companies found that competition from other railways, the preference of traders, and changes in the location of industry produced inevitable decline.

NOTES

1. F. Wishaw, *The Railways of Great Britain and Ireland* (1840), vi.
2. W. T. Jackman, *The Development of Transportation in Modern England* (1862), 665.
3. *Railway Times*, 29 October 1842.
4. C. MacLaren, *Railways Compared with Canals and Common Road* (1825), 82–9.
5. *Diary of H. Brown* 1828–9, 109–10. I am grateful to Dr C. Gulvin, Portsmouth College of Technology, for this reference.
6. A. C. O'Dell, *Railways and Geography* (1956), 57–9.
7. MacLaren, *op. cit.*
8. This point is dealt with more fully in my article 'Railways and the Transformation of the Scottish Economy', *Economic History Review*, XXIV no. 1 (February 1971).
9. R.C. *Canals and Waterways;* 1907 XXXIII, q. 29728, 31136–7; Appendix 39.
10. By 1870 almost every major Scottish town was served by at least two railway companies.
11. This underestimates the impact of the railways since many of them were constructed after 1834 and 1848, the base years for the calculations, and would be in competition with the turnpikes for only part of the period for which calculations were made.
12. This was not customary in England but lasted in Scotland until the mid-1840s when, after an attempt to introduce a clause into all Scottish railway bills making for full compensation for turnpike traffic losses, it was decided by the House of Commons that no claims for contingent losses could be sustained by either the subscribers or creditors of the road trusts.
13. S.C. *Turnpike Trusts;* 1839 XIX, iv.
14. *Railway Times*, 9 July 1842.
15. J. Mitchell, *Reminiscences of My Life in the Highlands*, vol. II (1883), 76.

16. *The Bailie*, 22 April 1896.   17. *Railway Chronicle*, 14 September 1850.
18. *Railway Times*, 31 August 1844.
19. Scottish Record Office (hereafter S.R.O.): *PYB(S)1/162*, pp. 5-9; *FCN4/1, passim. Herapath's Railway Journal*, 12 September 1854; 13 December 1865; 27 April 1867.
20. R.C. *Canals and Waterways, op. cit.*, Appendix 39.
21. S.C. *Railway and Canal Amalgamations;* 1846 XIII, 776-7.
22. *City of Edinburgh Archives*, Canals, misc., bundle 6.
23. In 1898 canals in Scotland belonging to railways made an average profit of £281 per mile compared to the £23 per mile of railway-owned canals in England. Independent canals in Scotland earned only £19 per mile; south of the border they obtained £259 per mile. *Parliamentary Papers*, 1900 LXXV, 5.
24. Apart from the canals which gained Acts unpublished work deposited in the National Library of Scotland, Edinburgh, shows over 200 small private canals constructed in Scotland. These were of short length and there is no evidence of the public being allowed to use them.
25. S.R.O.: *PYB(S)1/29, passim.*
26. *Aberdeen Journal*, 2 September 1845; *Railway Chronicle*, 8 December 1849; S.R.O.: *GNS1/1*, p. 457.
27. *Railway Chronicle*, 28 September 1850; S.R.O.: *NBR1/4*, 21 March 1950; *Railway Times*, 2 September 1843; *Railway News*, 5 April 1875; *Herapath's Railway Journal*, 3 May 1873.
28. *Herapath's Railway Journal*, 1 May 1852; 20 August 1853; 20 September 1857.
29. S.R.C.: *Carron Company*, GD58, sections 18/58, 18/78.
30. J. Lees, *A History of the County of Inverness* (1898), 347.
31. *Herapath's Railway Journal*, 8 May 1880.
32. The initial experiences of the Glasgow, Paisley & Greenock have been documented more fully in T. R. Gourvish, 'The Railways and Steamboat Competition in Early Victorian Britain', *Transport History*, IV no. 1 (February 1971).
33. See for example *Herapath's Railway Journal*, 14 September 1844; 20 March 1845; 9 June 1855.
34. S.C. *Railway Companies' Powers;* 1864 XI, iii-iv.
35. S.R.O.: *Report of the Investigation into the Affairs of the North British Railway*, 15 November 1866, 10.
36. S.R.O.: *HRP(S)35*, 6.
37. W. C. Galbraith, *The Caledonian Steam Packet Company Ltd.* (1949), 1-7.
38. S.C. *Railway Company Amalgamations;* 1872 XIII, xix. S.C. *Railway Companies' Powers, op. cit.*, 629.

3 Railroads and the American economy: the Fogel thesis in retrospect

D. L. LIGHTNER

Twenty years ago Robert Fogel startled historians by advancing the thesis that railroads had not been indispensable to American economic growth.[1] Almost as novel as the thesis itself was Fogel's argument and evidence on its behalf, for he was one of the pioneers of what has come to be called the new economic history, an approach characterised by quantification, model building, and the formal use of economic theory. *Railroads and American Economic Growth*, published in 1964, inaugurated a controversy that has since generated a large literature. Some scholars were so impressed by Fogel's work that they copied his methods and applied them to studies of the impact of railroads in many nations. Others were fiercely critical, attacking Fogel's evidence and rejecting his conclusions.[2] Fogel supplemented his pathbreaking book with several articles, in the most recent of which, 'Notes on the social saving controversy', he replies to many of his critics.[3] The time seems ripe for an interested outsider to try to assess the adequacy of Fogel's defence and to ponder how well the Fogel thesis holds up after two decades of debate. Because of the limitations of space, our discussion is confined strictly to the American case, cites only part of the relevant literature on many points, and ignores disputes within the ranks of the critics, who sometimes quarrelled with one another as much as with Fogel. We begin by briefly summarising the thesis and describing Fogel's evidence and methods. Next, we examine his comments on a variety of technical issues that were raised by critics who found fault with his data or techniques. Finally, we assess Fogel's response to some conceptual criticisms.

I

Fogel believed that the impact of railroads upon economic growth could be determined by measuring the social saving that resulted from shipping goods by the transport system that existed in the United States in 1890 rather than by a hypothetical 'next best' system that might have existed had there been no railroads.

The latter system would have relied primarily upon water transport, whereas in the real world of 1890 railroads were, of course, predominant.[4] Fogel decided initially to limit his focus to the movement of agricultural commodities. He divided that process into two aspects: first, the inter-regional shipment of surplus foodstuffs from the Mid-west to the East and South, and secondly the intra-regional shipment of more numerous agricultural commodities over shorter distances within the major sections of the country.[5] In calculating the cost of moving goods by the hypothetical non-rail alternative, Fogel took into account not only the obvious freight charges that would have been incurred but also certain indirect costs. For example, in order to allow for the disruption of water transport during the winter and for the relatively slow speed with which goods would have moved had boats and waggons replaced the rails, he included in his non-rail alternative the cost of carrying the additional inventories that would have been required to bridge the gaps in supply resulting from slow or interrupted water transport. Other indirect costs which he took into account were: the cost of cargo losses, which he estimated on the basis of insurance data; transhipping costs; costs of waggon haulage to markets not on waterways; and capital costs not reflected in water rates because of government subsidies.[6] In his hypothetical non-rail transport system Fogel presumed the existence of somewhat better roads and also of some 5,000 miles of canals that he thought were likely to have been built had there been no railroads in 1890.[7]

Using this model, Fogel carried out extensive calculations, drawing upon economic theory and statistical techniques to generate the necessary data. His labours eventually led him to conclude that the agricultural social saving from railroads in 1890 amounted to no more than 1·78 per cent of GNP.[8] Extrapolating from that result, he went on to assert that the social saving from railroads in the transportation of all commodities amounted to less than 5 per cent of GNP in the same year.[9] Fogel acknowledged that railroads brought about some additional social saving in indirect ways, such as by stimulating the iron and steel industry, but he argued that those effects were of minor significance.[10]

II

Considering the immense task that Fogel had set himself, it is hardly surprising that other scholars raised a variety of technical objections to his data and methods. For example, critics noticed that in computing the cost of shipping foodstuffs inter-regionally Fogel had assumed that it cost, on average, 0·139 cents per ton mile to move grain by water. That was the actual rate at which wheat was carried on the all-water route from Chicago to New York in 1890, and Fogel had decided on the basis of 'casual examination of the available data'[11] that it cost about the same to move all grains by all the various inter-regional water routes. That procedure was entirely too casual for Peter McClelland, who held that the 0·139 figure was not at all representative. McClelland pointed out that the average freight rate per ton mile on New York canals was 0·26 cents, while the rate for moving wheat from St Louis to New Orleans was 0·19 cents by barge and 0·27 cents by steamer.[12] In his 1979 article Fogel responds to this criticism by emphasising that the rates cited by McClelland are for distances considerably shorter than the average interregional haul, which Fogel estimated at 1,574 miles, and that therefore the existence of those higher rates does not prove his original 0·139 figure to be in error.[13] Fogel's reply is persuasive, and it seems clear that his original cost estimate was not nearly so far off the mark as McClelland believed. Because the validity of Fogel's cost estimate depends upon the accuracy of his length-of-haul estimate, it remains possible that his 0·139 cents per ton mile figure is too low, but the amount of such estimating error, if it exists at all, is likely to be slight.[14]

In using the 0·139 cents per ton mile figure to calculate shipping costs in the non-rail case, Fogel assumed that the greater volume of water shipping that would have existed had there been no railroads could have been accommodated at the same unit cost as the volume that actually existed in 1890; in other words, he assumed that the long-run marginal cost curve of water transport was constant or declining. Many critics questioned that assumption,[15] but in his 1979 article Fogel defends it vigorously, arguing that nineteenth-century experts all agreed that the unit cost of water transport decreased as tonnage, distance and facilities increased. Fogel supplements this qualitative evidence with quantitative data on canal cost functions.[16] While he is convincing in his argument that the marginal cost curve of water transport was downward-sloping, acceptance of that proposition does not necessarily prove that in the non-rail case all grain could have been transported in 1890 at a cost of 0·139 cents per ton mile. One critic, Gilbert Fite, questioned the practicality of Fogel's model on the grounds that some of the rivers that were used in his hypothetical enlarged system of water transport were 'mere streams that would hardly float a corncob in high water'.[17] In his 1979 article Fogel meets his objection by suggesting that 5,000 miles of canals could have been built along such rivers. Unlike the 5,000 miles of canals that were allowed for in his original formulation of the non-rail case, however, these additional 5,000 miles were not studied by nineteenth-century engineers. Without detailed knowledge of the rise and fall in elevation and the adequacy of the water supply, it cannot be known whether or not these canals were technologically feasible, much less what it would have cost to build and operate them. How they would have affected the marginal cost curve of water transport is, therefore, a moot point.[18]

Several of Fogel's critics raised a quite different challenge to his calculation of water shipment costs. These writers argued that if railroads had not existed, then canals would have used monopoly power to raise their rates above marginal costs.[19] To this criticism Fogel replies, first, that the matter is unimportant because canals account for only 5 per cent of total transport charges in the non-rail case, and second, that canals would not have used monopoly power to raise their tolls because more than 80 per cent of the tonnage shipped via canal in 1890 was carried on canals that were owned not by private firms but rather by state governments or by the federal government, and that historically those governments had used their monopoly power to lower canal tolls below marginal costs rather than to raise them above marginal costs.[20] Fogel's first point is unpersuasive, because one reason for the small share of canal charges in overall water transport costs in 1890 was the very fact that canals were not using monopoly power to raise their rates. Had they done so, their share of overall costs would have increased. (Their share would increase even more if the 5,000 miles of canals along 'doubtful' rivers were added on to the 5,000 miles of canals included in Fogel's original model of the non-rail case.) Fogel's second point also is vulnerable, because it involves the assumption that public ownership would have been as prevalent on the hypothetical canals that were included in his non-rail model as it was on the canals that actually existed in 1890. That assumption is inappropriate. In the United States a shift towards less public and more private ownership of transport facilities occurred at the same time as the canal era gave way to the railway age, but there was no causal relationship between those simultaneous developments. The complex shift in social values that led Americans to turn towards private ownership of railways could just as well have led them to turn towards private ownership of canals in the non-rail case, thus opening up much greater opportunity than Fogel acknowledges for the exercise of monopoly power to raise canal tolls above marginal costs.[21] When all is said and done, however, Fogel still appears to be the winner of this particular argument, because his critics have failed to distinguish between the equity and the efficiency effects of monopoly pricing. Fogel points out that if canals had used monopoly power to raise their rates above marginal costs, then their increased revenues would have been mainly an income transfer, which would not affect the social saving calculation.[22]

Some criticism was directed at Fogel's calculation of railroad shipping costs. Critics argued that railroad freight rates in 1890 were above marginal costs because of the exercise of monopoly power by railroads.[23] Fogel responds to this argument by computing for 1890 the ratio of the net earnings of American railroads to the reproduction cost of the capital embodied in them and showing that the resulting figure is below the 1890 yield on industrial common stocks, the 1890 yield on the stock of utilities, and the 1890 interest rate on commercial paper.[24] While this exercise does demonstrate that railroads in general were not reaping monopoly profits, it remains possible that the particular railroads whose rates figured in Fogel's social saving calculation did not conform to the national norm.[25] A possibly more serious flaw in his rail cost calculations is that he drew his rate data from tariffs filed with the Interstate Commerce Commission. To the extent that rebating took place, these public tariffs would overstate the marginal cost of shipping by rail. One prominent business historian branded Fogel's rail cost calculations 'ludicrous', claiming that 'in the world of 1890, a shipper had to be deaf, dumb, and blind to pay the railroad rate asked for in the printed tariff'. In his 1964 book Fogel had recognised this rebate problem, at least as it affected his inter-regional cost calculations, but in his 1979 article he does not deal with it.[26]

Fogel also takes another tack. He argues that even if his critics are right on some points, their criticisms nevertheless do not invalidate his conclusion that the agricultural social saving amounted to less than 1·78 per cent of GNP. Fogel points out that in computing shipping costs in the non-rail case he had assumed that exactly the same quantities of goods would have been shipped in exactly the same pattern as existed in the real world of 1890; he did not allow the non-rail economy to adjust to the greater transport costs by changing the mix of goods and the pattern of shipments. This unrealistic assumption that the demand for transport was completely inelastic introduced an upward bias into the social saving estimate.[27] Fogel does not pretend to know what the elasticity of demand for transport actually was in 1890, but he can and does calculate the amount of upward bias which various elasticities would introduce into his social saving estimate (see table 1). Fogel then

Table 1 *Estimate of the potential upward bias in Fogel's agricultural social saving estimate as a function of elasticity of demand for transport services*

Elasticity of demand	Potential upward bias (% of GNP)[a]
0·0	0·00
0·4	0·23
0·75	0·39
1·0	0·50
1·5	0·68
2·0	0·83

[a] Fogel expresses his estimates of upward bias as percentages of the 'true' social saving (i.e. the social saving without the upward bias due to the assumption that elasticity of demand was equal to zero). Because Fogel's original estimate of the social saving was 1·78 per cent GNP, the amounts of upward bias expressed as percentages of GNP can be calculated according to the formula

$$X = 1\cdot78 - \frac{1\cdot78}{1 + Y}$$

where X is the upward bias expressed as a percentage of GNP, and Y is the upward bias expressed as a percentage of the 'true' social saving. The results of such calculations appear here in column 2.

Source. Fogel, 'Notes', 10–12 and table 2.

presents a series of calculations designed to demonstrate that each of the possible downward biases cited by his critics would likely be overwhelmed by the upward bias arising from the assumption that the elasticity of demand for transport was equal to zero. Fogel's method here may be clarified by an example. It will be recalled that some critics said that he had erred in assuming that the water rate from Chicago to New York was representative of all inter-regional water rates. To demonstrate that any downward bias resulting from this factor cannot amount to much even if the critics were right, Fogel now calculated that even if he had used the St Louis to New Orleans barge rate instead, his social saving estimate would have risen by only 0·03 per cent of GNP.[28] The possible downward bias resulting from his original use of the Chicago to New York rate appears, therefore, to be trifling in comparison with the upward bias arising from the assumption that demand for transport was completely inelastic.

While Fogel's method here is ingenious, it is not unassailable. One weakness is obvious: since he does not know what the elasticity of demand for transport was in 1890, he cannot know how much upward bias results from the assumption that the elasticity was equal to zero. Fogel seemingly regards 0·4 as a lower bound on the actual elasticity of demand, but he provides no rationale for doing so. A second weakness also seems evident: even if every potential downward bias is, taken by itself, negligible compared with the upward bias of which Fogel makes so much, it remains conceivable that a congeries of small downward biases could, in total, equal or exceed the upward bias. As a matter of fact, if all the tiny percentages of GNP that result from Fogel's hypothetical calculations of potential downward biases are added up, the resultant total exceeds his estimate of the upward bias in the case of an elasticity of demand of 0·4.[29] This result certainly does not prove that his estimate of the agricultural social saving really contains a significant downward bias, for Fogel has presented convincing arguments against his critics on some of the points at issue. However, it does illustrate the possibility that a conglomeration of small downward biases could exceed the upward bias arising from the assumption that the demand for transport was inelastic. That there may be such a conglomeration of small downward biases in Fogel's social saving estimate seems possible not only because there are weaknesses in some of his replies to his critics but also because his critics have raised a variety of additional points which he has not addressed. To mention just a few examples: Patrick O'Brien questioned the accuracy of Fogel's estimates of the amounts of foodstuffs hauled inter-regionally;[30] Paul David claimed that by improperly assuming ubiquitous access to waterways Fogel had understated the amount of waggon haulage required in the non-rail case;[31] and E. H. Hunt asserted that Fogel had made insufficient allowances for cargo losses, slow transit and winter closures.[32]

Fogel also has failed to reply to some important technical criticisms that were aimed not at his computation of the agricultural social saving but rather at the procedure whereby he went on to estimate the social saving in the transport of all commodities. To obtain his estimate of the overall social saving, Fogel first pointed out that more than half of all non-agricultural freight consisted of minerals. Then, because the production of minerals was concentrated as to location and because minerals were cheap to store, he deduced that the social saving per ton mile on non-agricultural freight was likely to be lower than the saving on agricultural freight. Finally, he asserted that because agricultural products made up about 25 per cent of all freight in 1890, he could, after certain adjustments to prevent double counting, multiply by four his estimate of the agricultural social saving in order to obtain an estimate of the social saving on all commodities. Fogel recognised the crudity of this process of extrapolation but considered it adequate to demonstrate that the overall social saving from railroads in 1890 was below 5 per cent of GNP.[33] Some critics were not impressed. E. H. Hunt questioned whether the social saving on minerals would be as small as Fogel supposed. In calculating the agricultural social saving, Fogel had used relatively low waggon costs because the movement of farm produce made use of men, animals and equipment that would have been underemployed following the harvest, and because farmers usually carried purchases back home on the return trip. Hunt pointed out that neither of these cost-reducing factors would be present in the hauling of minerals by waggon. Hunt also wondered how minerals located in states like Arizona and Colorado could have been exploited without railways.[34] Raising a quite different challenge to Fogel's extrapolation procedure, Albert Fishlow suggested that agricultural goods constituted about 17 rather than 25 per cent of all freight.[35] Fishlow's assertion, if correct, is devastating to Fogel's estimate of the overall social saving, yet he has not replied to it.[36]

III

In addition to making technical criticisms of Fogel's methods and data, the critics raised a number of conceptual issues that are crucial to an assessment of the Fogel thesis. In his 1979 article Fogel replies convincingly to his critics on some of these questions,[37] but he leaves others unanswered.

Is 5 per cent of GNP a small social saving? Fogel tacitly assumed that it was, but many critics disagreed. Stanley Lebergott pronounced Fogel's finding 'startling, but uninteresting',[38] and Paul David labelled it 'jejune'.[39] The problem is that GNP is such a large figure (roughly $12 billion in 1890)[40] that almost anything is small by comparison. Indeed, the whole 'transportation and other public utilities' sector of the economy was itself only about 10·7 per cent of 1890 GNP.[41] If expressed not as percentages of GNP but rather in relation to transport costs, Fogel's social saving figures certainly seem far from insignificant. In *Railroads and American Economic Growth* Fogel himself observed that 'the absence of the railroad would have almost doubled the cost of shipping agricultural commodities interregionally' even though the inter-regional social saving on agricultural goods amounted to only 0·6 per cent of GNP.[42] Robert Thomas and Douglas Shetler pointed out that the social saving from railroads also looms large when compared with the social savings produced by other innovations of the nineteenth and twentieth centuries: so far as is known, nothing else approaches the 5 per cent saving attributed by Fogel to the railroad.[43] Critics also questioned the 'smallness' of the 5 per cent social saving figure on the grounds that it might have been considerably larger had Fogel focused his study upon the year 1900 rather than 1890. E. H. Hunt argued that, since some 73,000 miles of railroad were built in the United States in the 1880s, it was possible that a considerable portion of the railway network was too new to have attained maximum efficiency by 1890.[44] Louis Hacker, on the other hand, stressed the fact that during the 1890s wheat and livestock production shifted westward beyond the ninety-eighth meridian into an area where alternatives to the railroad were particularly inferior.[45] There seems to be no reason why Fogel could not have used data from the year 1900 for his calculations, as the impact of motor vehicles was still negligible at the turn of the century.[46]

A final objection to regarding the 5 per cent social saving as small is that to do so is to ignore the cumulative effect that such a social saving might produce over time. This effect may be illustrated by a hypothetical example (see table 2). Assume

Table 2 *Cumulative effect of railroad social saving on GNP over time, assuming GNP growth rate of 4 per cent per annum and railroad social saving of 5 per cent GNP (GNP with railroads = 100 in year 1)*

Year	Economy with railroads	Economy without railroads
1	100·0	95·0
2	104·0	98·6
3	108·2	102·4
4	112·5	106·2
5	117·0	110·3
6	121·7	114·5
7	126·5	118·8
8	131·6	123·3
9	136·9	128·0
10	142·3	132·9

Source. See text and note 47.

that GNP grows at 4 per cent per year and that the social saving from railroads is 5 per cent of GNP. Let the GNP of the economy with railroads be 100 in year 1. In the same year the GNP of the economy without railroads will be 5 per cent less, or 95. After one year the economy with railroads will have grown by 4 per cent. The economy without railroads will not have grown at the same rate, because in the non-rail case 5 per cent of every increase in output must be devoted to transport. The growth rate of the non-rail economy is therefore reduced to 3·8 per cent per annum. As time goes on the gap between the two economies will widen. After five years the economy with railroads will be 5·7 per cent larger than the economy without railroads, and after ten years it will be 6·6 per cent larger.[47] Fogel has not responded to this argument or to any of the other arguments raised by those critics who believe that his 5 per cent social saving figure should be regarded as significant. Instead, in his 1979 article he continues to cling to his tacit assumption that a 5 per cent railroad social saving is small.[48]

Another body of criticism has centred on the conceptual adequacy of Fogel's model. Critics argued that the absence of railroads would have affected GNP in ways not captured by his measure of the social saving. For example, Paul David said that, by lowering marginal costs in industries that used transport, railroads allowed such industries to achieve economies of scale.[49] Fogel replies by acknowledging the theoretical validity of David's point but insisting that, given what we know about scale economies in the nineteenth century, the impact of this factor is not likely to be significant.[50] Several writers wondered at Fogel's omission of any consideration of railroad passenger services. Because passenger fares on boats were generally lower than railroad fares, Fogel perhaps felt justified in assuming that the loss of railroad passenger services would not have reduced GNP. That reasoning is invalid, however, if indirect costs are taken into account. J. Hayden Boyd and Gary M. Walton estimated that passengers in 1890 would have valued the time that they saved travelling by rail at about 2·6 per cent of GNP.[51] Because much of this saving came in the form of leisure time, it would be wrong to assert that GNP would actually have fallen by 2·6 per cent had there been no railroads, but clearly there would have been some reduction.[52] There was, moreover, a fair amount of passenger travel for which railroads were virtually indispensable. 'Over what route,' asked George Rogers Taylor, 'would persons have travelled between Boston and Minneapolis in 1890 when northern waterways were obstructed by ice?'[53] Fogel has not responded to this criticism.[54] It is easy to speculate upon other ways in which the absence of railroads might have retarded economic growth. Would foreign investment have continued to flow into a water-based American transport system after the financial failure of many canal companies in the 1840s?[55] Could the crucial role of the railroads in developing management techniques and stimulating the machine tool industry have been duplicated by other means in a non-rail world?[56] How would the livestock and meat-packing industries have been affected by the absence of the refrigerated railroad car?[57] Could John D. Rockefeller have built Standard Oil without taking advantage of railroad rebates?[58] Finally, we may ask whether an America deprived of locomotives could nonetheless have had steamboats. The two innovations are so closely related technologically that it seems inconceivable that one could exist for long without the other being invented. But if the alternative to an America with railroads was an America without either railroads or steamboats, then the social saving was far higher than the 5 per cent calculated by Fogel.[59]

In his 1979 article Fogel concedes that his model is not complete,[60] that 'the influence of railroads on economic growth is too complicated to be encompassed by one or even a few models',[61] and even that 'it is likely that some of the most important aspects of the connection between railroads and economic growth will not yield to formal modelling and will be better served by more traditional historiographic approaches'.⁶² Yet Fogel seems oblivious to the consequences of these admissions. He concludes his article by stressing that scientific creations 'approach perfection quite gradually' and asserting that the controversy over his thesis is a sign that 'the method is working'.⁶³ If these closing remarks are intended to mean only that the debate over the thesis has led to much technical improvement in Fogel's analysis, then they are unexceptionable. If they are intended to imply that the quantitative approach is leading the way to some brave new world of objective historical knowledge, however, then they must be challenged. The Fogel thesis not only rests upon many assumptions, but also reflects a particular point of view. To compare an economy with railroads to an imaginary economy without them is a methodological approach that is congenial to neoclassical economists, who are accustomed to measuring the opportunity costs of various alternatives. This same approach can contribute to historical knowledge, because the study of past economic choices is one aspect of economic history.⁶⁴ But it is not the whole of it. Historians are concerned also with providing an ordered narrative of past events, and for that purpose the social saving approach is unsuited. As Fogel himself says, 'the new economic history has done little to change our perception of the sequence of events that constitute the history of modern transportation'.⁶⁵ Historians do not usually assess the significance of phenomena by considering alternatives; they would not say that John Wilkes Booth's assassination of Abraham Lincoln was insignificant because Lincoln could have been killed in many other ways or by many other people. From this perspective, the Fogel thesis is not necessarily wrong, but it is irrelevant; to demonstrate that many of the economic changes wrought by the railroad could have been achieved by other means does not alter the fact that the railroads did what they did.⁶⁶ No historian will ever again say that railroads were indispensable to American economic growth⁶⁷ — in that narrow sense the Fogel thesis is triumphant — but there is no reason for historians otherwise to alter their customary practice of assigning to the railroad a central role in any narrative account of American economic history.

IV

How fares the Fogel thesis after twenty years? The answer is mixed. With respect to the author's data and techniques, *Railroads and American Economic Growth* has proved to be more sound than many readers, including this one, once supposed. While it is possible that the collective impact of many small downward biases could outweigh the upward bias arising from the assumption that demand for transport was inelastic, Fogel's estimate of the upper bound on the agricultural social saving is better grounded than his critics had supposed. Fogel has demonstrated that a number of what had seemed to be major technical flaws in his calculations either do not exist at all or else have only a negligible potential impact. The most serious technical weakness in the Fogel thesis is the crude procedure whereby the estimate of overall social saving was extrapolated from the agricultural social saving. Unless and until the social saving on non-agricultural commodities is calculated with the same care and precision that Fogel lavished upon agricultural goods, those of us who choose to remain agnostic on the question of whether or not the overall social saving was below 5 per cent of GNP will continue to have reasonable grounds for refusing to embrace the faith. With respect to conceptual issues, the Fogel thesis has fared less well. Its author has chosen to ignore a variety of objections which undermine not so much the validity as the utility of his enterprise. To know that the social saving from railroads amounted to less than 5 per cent of GNP is not to know very much, once it is realised that there is no pre-emptive basis for regarding a 5 per cent saving as small and that the model from which that estimate is derived is incomplete, viewing the significance of railroads from a perspective that is alien to the norms of historical scholarship. Because of its conceptual weaknesses, the Fogel thesis has not provided and cannot provide a definitive measurement of the contribution of railroads to American economic growth. However, it has led to an improved understanding of some aspects of the contribution made by railroads, to a more precise and meaningful discussion of many questions that remained unresolved, and to a growing appreciation by both historians and economists of what the new economic history can do — and of what it cannot do.

Notes

1 Some aspects were outlined in R. W. Fogel, 'A quantitative approach to the study of railroads in American economic growth: a report of some preliminary findings', *Journal of Economic History*, XXII (1962), 163–97, and the thesis was stated in full in R. W. Fogel, *Railroads and American Economic Growth: Essays in Econometric History* (Baltimore, 1964).

2 For an extensive bibliography see P. O'Brien, *The New Economic History of the Railways* (New York, 1977), pp. 115–18.

3 R. W. Fogel, 'Notes on the social saving controversy', *Journal of Economic History*, XXXIX (1979), 1–54. O'Brien's bibliography may be supplemented by the more recent works cited in this article.

4 Fogel, *Railroads*, pp. 10–20, 207–9.

5 Ibid., chs. 2–3.

6 Ibid., pp. 41–8, 85–91.

7 Ibid., pp. 75–110.

8 Ibid., pp. 47, 110, 219.

9 Ibid., p. 223.

10 Fogel calculated that railroads absorbed only 17 per cent of total iron production in the period 1840–60; *ibid*., p. 134. In a later article Fogel acknowledged that in the post-Civil War period railroads absorbed a large proportion of total steel production (between 50 and 87 per cent of steel output during each of the years between 1871 and 1890), but he argued that non-rail demand for steel was growing so rapidly that, even without the railroads, total steel production would have lagged behind the figures actually attained by only six years. R. W. Fogel, 'Railroads as an analogy to the space effort: some economic aspects', *Economic Journal*, LXXVI (1966), 16–43, reprinted in P. Temin, ed., *New Economic History: Selected Readings* (Harmondsworth, 1973), pp. 239–40.

11 Fogel, *Railroads*, p. 38.

12 P. D. McClelland, 'Railroads, American growth, and the new economic history: a critique', *Journal of Economic History*, XXVIII (1968), pp. 105–6. Harry Scheiber raised the same point, citing rates in Ohio and elsewhere c. 1850. H. N. Scheiber, 'On the new economic history — and its limitations: a review essay', *Agricultural History*, XLI (1967), 387–8.

13 Fogel, 'Notes', 15.

14 In his 1979 article Fogel presents a graph based on data pertaining to the upper Mississippi river and showing how freight rates declined as length of haul increased. Superimposed on this graph are the various freight rates cited by Fogel's critics, and it is shown that the latter cluster tightly around the regression curve. Although this demonstration of the relationship between costs and distance is impressive, it does not alter the fact that Fogel's 0·139 cost figure is accurate only if his estimate of the average length

of haul is correct. If the latter was anything less than 1,574 miles, then the 0·139 cost estimate is too low. There appears to be room for estimating error in the 1,574 figure, which Fogel obtained by drawing a random sample of thirty routes from a population of 825 routes linking pairs of cities. Fogel, *Railroads*, p. 40.

15 For example: P. A. David, 'Transport innovation and economic growth: Professor Fogel on and off the rails', *Economic History Review*, XXII (1969), 506–25, reprinted in Temin, *New Economic History*, p. 268; D. Wellington, 'The case of the superfluous railroads: a look at changing transportation patterns', *Economic and Business Bulletin*, XXII (1969), p. 34; McClelland, 'Railroads, American growth', 115, 119.

16 Fogel, 'Notes', 23–9.

17 G. C. Fite, review of *Railroads and American Economic Growth*, in *Agricultural History*, XL (1966), 148.

18 Fogel makes the unfounded assumption that the construction cost of these canals would be identical to that of the canals that he allowed for in his original model. Fogel, 'Notes', 19 n.

19 For example: L. M. Hacker, 'The new revolution in economic history: a review article based on *Railroads and American Economic Growth* . . .', *Explorations in Entrepreneurial History*, second series, III (1966), 166; E. H. Hunt, 'The new economic history: Professor Fogel's study of American railways', *History*, LIII (1968), 15; C. M. White, 'The concept of social saving in theory and practice', *Economic History Review*, second series, XXIX (1976), 92.

20 Fogel, 'Notes', 19–21.

21 That there is no necessary connection between the form of transport technology and the pattern of ownership is obvious in view of the great variation among nations on the degree of public versus private involvement in canal and railroad building. See the old but still useful descriptions in J. H. Clapham, *The Economic Development of France and Germany, 1815–1914* (4th edn, Cambridge, 1936), chs. 5, 7, 12. For example, Clapham remarks that when Belgium opted for government development, 'partly for the glory of the young state, partly because the government was resolute that the whole work must be carried out systematically', she was 'ahead of England in that she had a railway policy, when England was fumbling for a policy which she never found', *Ibid.*, pp. 140–1.

22 Fogel, 'Notes', 19 n, 23. Some critics also challenged Fogel's calculation of water shipment costs on the grounds that he had made insufficient allowance for government subsidies for canals and for river and harbour improvements. Fogel's defence of his original allowance for these uncompensated capital costs is brief but convincing. *Ibid.*, 16–17.

23 For example: S. Lebergott, 'United States transport advance and externalities', *Journal of Economic History*, XXVI (1966), 439; McClelland, 'Railroads, American growth', 114–15; P. D. McClelland, 'Social rates of return on American railroads in the nineteenth century', *Economic History Review*, second series, XXV (1972), 476–8; O'Brien, *New Economic History of Railways*, p. 44.

24 Fogel, 'Notes', 35–6.

25 Fogel based his interregional railroad rate for grain on the rate for shipping wheat by rail from Chicago to New York. For his intraregional calculation Fogel drew upon the local tariffs of five railroads in the North-east. Fogel, *Railroads*, pp. 38, 69–70, 85 n.

26 A. Martin, review of *Historian's Fallacies*, by D. H. Fischer, in *Journal of Economic History*, XXXII (1972), 970. Fogel, *Railroads*, pp. 36–7, 69.

27 Fogel, *Railroads*, pp. 19–20, 52.

28 Fogel, 'Notes', 16 n.

29 Fogel's illustrative calculations concern downward bias that might result from underestimation of the inter-regional water rate (bias equal to 0·03 per cent GNP), overestimation of railroad shipping costs because of monopoly profits by railroads (0·08 per cent GNP), underestimation of canal construction costs (0·10 per cent GNP), and omission of construction costs for canals along 'doubtful' rivers (0·10 per cent GNP). The total of these hypothetical downward biases (0·31 per cent GNP) exceeds the potential upward bias if elasticity of demand was 0·4 (bias equal to 0·23 per cent GNP—see table 1). Fogel, 'Notes', 16, 19, 36. Fogel makes similar calculations regarding two other possible sources of downward bias (underestimation of intra-regional water rate; underestimation of water rates due to exercise of monopoly power by canals) but presents the results in terms of his preliminary rather than his final estimate of the agricultural social saving. *Ibid.*, 16, 19. Finally, Fogel presents two more calculations, but these relate to possible railroad-induced economies of scale, a matter not originally included in his social saving model. *Ibid.*, 39–40.

30 Fogel built up his estimate of the quantity of foodstuffs shipped inter-regionally partly by calculating the amounts consumed in each region. O'Brien points out that Fogel used national *per capita* consumption data, making no allowance for regional variations in incomes and spending patterns. O'Brien admits, however, that the direction of any bias that might result from this factor is uncertain. O'Brien, *New Economic History of Railways*, p. 104, and cf. Fogel, *Railroads*, pp. 32–3.

31 'Fogel's calculations . . . implicitly assume that all sites along navigable waterways in the US would have been feasible and more or less equally efficient cargo trans-shipping points, and in that respect would have resembled sites along most railroad lines. In the absence of any allowance for additional capital outlays to create passable ubiquitous access to lake shores and river banks for both wagons and vessels, this simplification must lead to an understatement of the supplementary cost of intraregional wagon-haulage.' David, 'Transport innovation', in Temin, *New Economic History*, p. 269.

32 Hunt says that Fogel's use of insurance data to

compute the cost of losses in transit 'overlooks the uncertainty, inconvenience and delay occasioned by loss in transit — costs not covered by insurance premiums', while Fogel's use of inventory costs to quantify the effects of slow movement and winter closures is inapplicable to perishable goods and fails to take into account the seasonal unemployment of men and resources that would result from the winter closure of waterways. Hunt, 'New economic history', 14. Paul David echoes the latter criticism, noting that the cost of holding inventories accounted for 65 per cent of Fogel's inter-regional social saving estimate. David goes on to illustrate the magnitude of the storage problem by calculating that the Chicago stockyards would have to have occupied 10,000 acres, or half of all the privately utilized land in the city in 1890, in order to accommodate the necessary inventory of livestock. Although David's heuristic calculation appears impressive, it breaks down upon close examination. For one thing, David himself acknowledges that railroads occupied 4,500 acres of Chicago land, which would have been available for other uses had railroads been absent. More important, there is no reason to suppose that the required inventory of livestock would have been concentrated in Chicago; instead it could have been held in scattered locations in both the Mid-west and the East, the latter notion being implicit in Fogel's discussion of meat storage. David's calculation is cited by Alfred Chandler as a major reason for rejecting the Fogel thesis. David, 'Transport innovation', in Temin, *New Economic History*, pp. 269–71; Fogel, *Railroads*, p. 46 n; A. D. Chandler, Jr., *The Visible Hand: The Managerial Revolution in American Business* (Cambridge, Mass., 1977), p. 531 n.

33 Fogel, *Railroads*, pp. 219–23.

34 Ibid., p. 109. Hunt, 'New economic history', 10.

35 Fishlow also thought Fogel had overcompensated for double counting. A. Fishlow, *American Railroads and the Transformation of the Ante-bellum Economy* (Cambridge, Mass., 1965), pp. 59–61. Patrick O'Brien also expressed scepticism about Fogel's extrapolation, noting that agricultural goods and minerals combined accounted for only 62·5 per cent of all railroad ton miles in 1890. O'Brien, *New Economic History of Railways*, p. 112.

36 In his 1979 article Fogel says nothing about this subject. In his 1967 article he did not respond to Fishlow's criticism, but he drew upon Fishlow's book to devise a different method of extrapolation, by which he estimated that the overall social saving from railroads was only 3 per cent of GNP. Fogel, 'Railroads as an analogy', in Temin, *New Economic History*, p. 255.

37 Fogel replies to suggestions that what he had called social saving was merely 'a measure of estimating error' (Lebergott, 'United States transport', 439), that the benefit from railroads should be measured on the basis of 'best practice level of technology' on both canals and railroads (ibid., 441), and that the significance of railroads in economic growth should be determined by comparing the marginal social return on capital invested in railroads with the marginal social return on other investments (M. Nerlove, 'Railroads and American economic growth', *Journal of Economic History*, XXVI, 1966, 112). Because I consider these issues to be somewhat tangential and because I find Fogel's counter-arguments convincing, I omit discussion of these disputes. See Fogel, 'Notes', 8–10, 12–13, 29–34, for citations to additional critics and for his own replies.

38 Lebergott, 'United States transport', 438.

39 David, 'Transport innovation', in Temin, *New Economic History*, p. 282, italicises this word and spells it 'jejeune'.

40 This was the GNP estimate used by Fogel: *Railroads*, p. 41.

41 US Bureau of the Census, *Historical Statistics of the United States: Colonial Times to 1970* (Washington, 1976), p. 238.

42 Fogel, *Railroads*, p. 211.

43 Thomas and Shetler compared Fogel's estimate of the social saving from railroads (5 per cent GNP) with the social savings that had been calculated in studies of four other innovations: red fife wheat and the chilled steel plough in Canada (1·96 per cent GNP), the steamship in American international trade (0·25 per cent GNP), agricultural research in the United States (0·29 to 1·6 per cent GNP) and hybrid seed corn in the United States (0·06 per cent GNP). The authors cautioned, however, that comparing Fogel's figure with but four other estimates is like 'deciding an election on the returns of the first four precincts counted', R. P. Thomas and D. D. Shetler, 'Railroad social saving: comment', *American Economic Review*, LVIII (1968), 186–9.

44 Hunt, 'New economic history', 11.

45 Hacker, 'New revolution', 167–8.

46 In 1900 there were 8,000 automobiles registered in the few states then requiring registration, and 4,100 vehicles were sold; *Historical Statistics to 1970*, p. 716. Fogel, *Railroads*, p. 75 n, says that motor vehicles were 'a minor factor in transportation' as late as 1911–15.

47 The hypothetical nature of this exercise must be emphasised. Temin, *New Economic History*, pp. 14–15, points out that it is dependent upon the assumption that the absence of railroads represented a sustained reduction in the rate of technological change. The possible cumulative effect of the social saving appears to have been pointed out first by Folke Dovring at the 1966 meeting of the Economic History Association. See Scheiber, 'On the new', 389. My hypothetical illustration follows that of P. R. P. Coelho, 'Railroad social saving in nineteenth century America: comment', *American Economic Review*, LVIII (1968), 184–6, correcting an earlier effort by E. H. Hunt, 'Railroad social saving in nineteenth century America', *American Economic Review*, LVII (1967), 909–10. For a more elaborate illustra-

tion of the same point see O'Brien, *New Economic History of the Railways*, pp. 35–9. M. Desai, 'Some issues in econometric history', *Economic History Review*, XXI (1968), 6–7, argues that neoclassical economic theory is ill adapted to the study of dynamic situations. P. Temin, 'In pursuit of the exact', *Times Literary Supplement*, 28 July 1966, 652–3, stresses the index number problem, which becomes more troublesome as changes in GNP are charted over longer periods of time. For a good example of the index number problem see Fogel's critique of J. G. Williamson, *Late Nineteenth-century American Development: a General Equilibrium History* (Cambridge, Mass., 1974), in Fogel, 'Notes', 46, 47 n.

48 As is implicit in his remark, 'In recent years, I have often been asked why anyone should have expected a large social saving.' *Ibid.*, 48.

49 David, 'Transport innovation', in Temin, *New Economic History*, pp. 273–80.

50 Fogel points out that there is no evidence of economies of scale in agriculture and that many scale economies elsewhere were internal to the firm. 'Notes', 39–44.

51 The authors base their calculation upon the assumption that rail travellers in 1890 were analogous to modern automobile travellers, who place a monetary value upon their travel time that is approximately equal to the prevailing wage in manufacturing. J. H. Boyd and G. M. Walton, 'The social savings from nineteenth-century passenger services', *Explorations in Economic History*, IX (1972), 240, 245.

52 Although Boyd and Walton rightly insist that gains in leisure time were real gains and should not be ignored, they acknowledge that leisure time is excluded from conventional GNP measurements. Gains in working time, would, however, affect GNP. Passengers were also likely to have placed some value on the increased safety and comfort of rail travel, but those gains were not quantified. *Ibid.*, 239–40. The services of railroads in getting workers more rapidly to where they were needed must certainly also have affected GNP.

53 G. R. Taylor, review of *Railroads and American Economic Growth*, in *American Economic Review*, LV (1965), 891.

54 In discussing elasticity of demand for transport, Fogel mentions Boyd and Walton's study of passenger services; however, Fogel ignores that study's criticism of his own thesis: 'Notes', 10.

55 Lebergott, 'United States transport', 441, mentions the possible impact of 'the Pennsylvania debacle of the 1840s'.

56 Chandler, *Visible Hand*, ch. 3; T. C. Cochran, *Frontiers of Change: Early Industrialism in America* (New York, 1981), pp. 94–6; see also the works cited in Fogel, 'Notes', 39 n.

57 M. Y. Kujovich, 'The refrigerator car and the growth of the American dressed beef industry', *Business History Review*, XLIV (1970), 460–82. M. Yeager, *Competition and Regulation: the Development of Oligopoly in the Meat Packing Industry* (Greenwich, Conn., 1981), chs. 1–3.

58 R. W. and M. E. Hidy, *Pioneering in Big Business, 1882–1911* (New York, 1955), pp. 118–20, 197–8.

59 While no critic makes precisely this point, E. H. Hunt asks 'whether no railways implies no steamships', which 'would have influenced migration'; Patrick O'Brien says, 'it seems just as reasonable to define innovation in transport in the nineteenth century as the application of steam propulsion to carriers on railed-ways and waterways'; and Fritz Redlich argues that 'once the atmospheric engine had been developed into an efficient steam engine and the steam engine had successfully been put into boats, making steamers out of them, it was only a question of when the steam engine would be put on wheels, particularly as the railroad minus locomotive had existed for a long time'. Hunt, 'New economic history', 16; O'Brien, *New Economic History of Railways*, p. 29; F. Redlich, ' "New" and traditional approaches to economic history and their interdependence', *Journal of Economic History*, XXV (1965), 480–95, reprinted in G. D. Nash, ed., *Issues in American Economic History: Selected Readings* (3rd edn, Lexington, Mass., 1980), p. 8. Some indication of the significance of the steamboat in water transport may be gained from the fact that, according to Fogel, 90 per cent of waterborne freight service in 1890 was on natural waterways, and on such waterways the ton-mile rate for shipping agricultural products was less than half the comparable rate on canals. Fogel, 'Notes', 17 and 20, table 3. See also E. F. Haites and J. Mak, 'Social savings due to western river steamboats', *Research in Economic History*, III (1978), 263–304.

60 Fogel, 'Notes', 5.

61 *Ibid.*, 48.

62 *Ibid.*

63 *Ibid.*, 51–2. Fogel distinguishes between 'certain types of histories', which he equates with paintings, concertos and novels, and his own 'scientific' approach. The implication seems to be that traditional history possesses only an aesthetic value, while quantitative history builds towards objective truth. The implied dichotomy is false. See A. Rutten, 'But it will never be science, either', *Journal of Economic History*, XL (1980), 137–42.

64 Redlich, ' "New" and traditional approaches', in Nash, *Issues*, pp. 8–9, argues that because Fogel constructed a hypothetical model he did not produce history. The easy reply to Redlich is that, while the hypothetical model is not itself history, if it helps to illuminate history, then there is no reason not to use it as a tool.

65 Fogel, 'Notes', 48.

66 J. D. Gould, 'Hypothetical history', *Economic History Review*, XXII (1969), 206–7, stresses the difference in perspective between economists and historians. S. Fenoaltea, 'The discipline and they: notes on counterfactual methodology and the "new"

economic history', *Journal of European Economic History*, II (1973), 731–2, notes that historians seek sufficient rather than necessary causes. Cf. Fogel, *Railroads*, pp. 15, 235, taking the issue to be whether or not the railroad was a necessary condition.

67 'The new-style economic history championed by Fogel shows scant charity towards such lapses from analytical precision, however compellingly they may be used in the telling of true stories.' David, 'Transport innovation', in Temin, *New Economic History*, p. 263.

4 Railways and land values

J. J. PINCUS

Whether they were private land-grant lines as in the United States, or publicly owned systems as in Australia, the financing of railways and their rate policies have been sources of scandal and political conflict. After tentative support of private initiatives, colonial governments in the 1850s somewhat reluctantly accepted responsibility for the construction and operation of most Australian railway lines. Initially a few British technical experts, guided by public investigations, determined routes, durability and gauges (although legislatures controlled the speed with which construction monies were spent). By the 1870s, however, as politics grew to be more dominant, planning became less public. Parliamentary representatives scrambled to promote local interests. Railways presented a chronic political problem which repeated changes in administration, and even the much-noted invention of the statutory corporation, did not solve. The sequence, from a selective policy implemented by independent technical experts or commissions to the spoils system or 'log-rolling,' echoes that ascribed by H. N. Scheiber to Ohio canal-building in the antebellum period.[1] The results in the nineteenth century were, according to N. G. Butlin, 'a heavy overseas debt burden and a cumbersome, inefficient and wasteful system of communications'. Assessments of Australian railway investment and performance in the present century have not been much more flattering.[2]

That the defect lay in state ownership in Australia was suggested, for example, by E. A. Pratt and F. W. Eggleston.[3] However, the alternative institution, private railways with land grants, was not, at least in America, an undisputed success. In popular culture, railway promoters were robber barons who made fortunes by corrupt and shady means. Warning that the final evaluation of the policy has not been made, S. L. Engerman's careful assessment seems to be that, although the railways themselves were economically desirable, the land grants and other government aids were excessive, representing an unnecessary transfer from the general taxpayer to railway investors.[4] There is also evidence that settlement was delayed by the withdrawal of large sections from the public domain: nineteenth-century promoters of

American railways, especially those receiving the later land grants, have been accused of making their money on land speculation, and of not being as interested in running lines as in profiting from land. In addition, fierce political battles raged between railways and their clients both before and after the advent of government regulatory agencies.

The difficulties were different in the two countries, but the causes were similar. This article discusses the heart of the rural railway problem — the creation of land values — and some of the effects of the struggle over their disposition among landholders, railway entrepreneurs or departments, and governments. Simple models of railway economics (equally applicable to other public facilities like urban rapid transit systems, highways, port works and water supply) display the problem of public choice between state ownership and private land grants. Both institutions created economic and political problems due to the difficulty of making the appropriate 'earmarking', 'typing' or 'dedication' of the net economic benefits of railways. This can be explained as follows.

By the use of multi-part pricing, price discrimination or other means, a monopoly supplier may capture some of what would, in a competitive market, accrue to customers as 'consumer surplus'.[5] In the railway case this capturable surplus is manifest as increases in land values. It will be shown that, when railway entrepreneurs take land values fully into account, they have an incentive to make economically efficient decisions about investment, pricing and output. Although full land grants secure this desirable outcome, they also concentrate benefits totally in the hands of entrepreneurs. Partial land grants can thus be regarded as a compromise between efficiency and equity. Whether full or partial, once the grants were sold and the lines operating, railways had an incentive to raise rates, undermining the financial expectations held by farmers when they bought land. Land sales 'untied' the grants. Shippers, once lines were operating, lobbied for rate reductions of a magnitude not foreseen by railway entrepreneurs when planning investments. Therefore, even though partial land grants may have promised relative economic efficiency with dispersed benefits, they nevertheless gave rise to some violation of contracts, implicit or explicit, between governments and rail entrepreneurs, and between railways and settlers.

Towards a solution of the public choice dilemma state ownership of railways could contribute nothing unless, once again, net economic benefits were appropriately 'earmarked'. The mechanisms developed in Australia for this purpose were complicated by a number of factors. They include the imperfect link between the finances of state railways and those of the state departments collecting land rents. Moreover, state railways were not run for profit, were exposed to political interference and were prone to uneconomic expansionism. The struggle over land values resulted in the dissipation of economic benefits through over-branching, through freight charges being unrelated to cost, and through unwise 'developmental' public works like irrigation.

I. Demand for rail freight services

We begin by establishing connections between land values and the cost of two complementary means of carriage: animal-powered waggon for local transport, and rail for reaching the distant, final market. In fig. 1 the cost of waggon cartage is assumed to have been a constant per ton mile, t, equal to the slope of OT.[6] A farm at distance

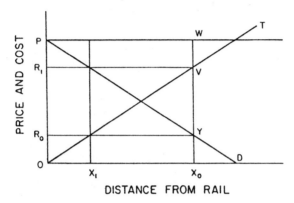

Fig. 1 Demand for rail services

X_0 from the railhead thus incurred tX_0 (= VX_0) in transporting one ton to rail. For simplicity, assume that all farmers received the same excess, P, of railhead price over production costs. The maximum that a farmer at location X_0 would have been willing to pay for rail freight per ton, then, is the surplus of the product price over the costs of producing and carting the output to rail; that is, $P - t X_0$ (= $WV = YX_0$). The line PYD, formed by 'reflection' of OT upon the horizontal through P, shows these surpluses for all locations within the rail catchment.

We are interested in the demand for the services of a railway *line* which passed through a considerable number of railheads. In fig. 2 we have drawn the ordinary demand curve for rail, derived from fig. 1 by transforming the abscissa according to the relationship between the quantity of produce sent to rail and the distance from the railhead that the product was grown. This transformation is discussed in note 7. Without specification of a full model, it is not clear how the various factors discussed in that note balance, and so we have used a simple linear form in fig. 2 for the demand for rail services offered by a railway containing a number of railheads.

II. Changes in land values

Figures 1 and 2 have a familiar, Ricardian use in that we can read off the difference in annual rents commanded by two farms that were identical except for location. Say the rail freight rate was R_0, so farms beyond X_0 did not find it profitable to

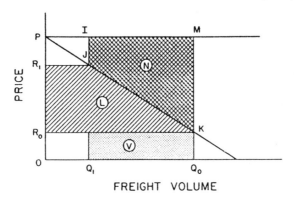

Fig. 2 Demand for rail services

cart produce to rail, i.e. costs of production plus all transport charges from beyond X_0 exceeded the output price. Location X_1, however, would have commanded a rent differential over X_0 (and farms farther away) equal to the cost of waggon transport from X_0 to X_1. In fig. 1 this rent differential is equal to $R_1 - R_0$, that is, equal to the rise in rail freight sufficient to shift the margin of rail catchment from location X_0 to location X_1.

Turning now to fig. 2, by extension of the argument a *fall* in the rail freight rate from R_1 to R_0 caused an increase in aggregate land value equal to the hatched area $R_1 JKR_0$, that is, by the increase in the 'consumer surplus triangle'. Call that increase L. (If no rail existed prior to the setting of freight rate R_0, then L would be equal to the whole triangle, i.e. the sum of the single-hatched portion and area PJR_1.) When the rail rate fell from R_1 to R_0 an extra quantity of produce equal to $Q_1 - Q_0$ was carted to the railhead, at a waggon cost, N, shown by the cross-hatched area IMKJ. By construction, $L = N$: the increase in land value equalled the increase in animal-powered waggon costs induced by the fall in rail freight rates. This allows us to place an upper limit on the increase in land value: in the linear case it can never exceed half the net value of the increase in product carried by rail; net, that is, of costs of production.[8]

Who retained the land value increase depended on whether the same freight rate (say, R_0) was charged to all farmers — if so, the landowners received the increase — or whether farms closest to the railhead were charged more than those farther away. At the extreme, a railway that was able to discriminate perfectly between farmers according to their maximum willingness to pay for rail freight (as shown by the demand curve PYD) would capture all the increase for itself. Although some freight-rate discrimination was practised, via charges for carriage on branch lines and in other ways, perfect discrimination was almost impossible, because it required knowledge of the origin of freight offered for carriage. Discrimination in the land

III. Value of a railway franchise

What subsidy would induce an entrepreneur to operate a rail monopoly in a particular area, or, if the proposed line were profitable, what premium would an entrepreneur pay? The answer depends not only on the relation of costs to demand but also on the pricing strategy that the monopolist could pursue. In particular, the minimum subsidy would be required or the maximum premium elicited — outcomes most favourable for the granter of the franchise — when the monopoly was free to discriminate in pricing, either directly through freight rate variations or indirectly through land rental variations. Without perfect price discrimination, rail entrepreneurs would be willing to finance only some of the economically desirable lines and, for the lines that were built, would wish to charge too much for freight services.[9]

Before a line was built, all costs were variable. To adopt the simplest case, let cost have two components, sunk and variable. The former consisted of the capital costs of construction (including any land purchased) and rolling stock, which, when converted to an annual charge over the life of the line, had a value equal to K.[10] This annual 'fixed' cost, invariant to the quantity of freight carried, is assumed to include a normal return to capital tied up. Operating costs are taken to have been a constant amount per unit of freight carried, and are represented by the 'marginal cost' line, mc, in fig. 3. For convenience, cost K is represented by the rectangular area under the rectangular hyperbola, drawn using mc and the freight rate ordinate as axes, so that ac shows average total costs for any quantity of freight.

Without a subsidy no entrepreneur would build and operate a line that was not expected to cover all costs. Consider first a line with neither land grants nor the ability to charge different freight rates to different farmers. For a break-even result or better, annual revenue must equal or exceed annual cost, so that the demand curve, BFD, must touch or cut the ac curve. This is not the case in fig. 3. The best that a potential monopolist could do would be to operate where losses are minimised, which is where marginal revenue, mr, cuts mc. Freight rate would be R_1, and freight quantity Q_1. The annual loss would be the shaded rectangle, representing the smallest annual payment that a non-discriminating monopolist would accept as subsidy to build and operate the line.[11] Land values would be higher with the railway than without, by an amount shown as the triangle R_1 BF. The figure has been drawn so that, with R_1 as freight rate, the railway generated more costs than benefits. That is to say, a fully integrated firm that bore all the costs and enjoyed all the benefits (including raised land values) would not build and operate the line at freight rate R_1.

Rather than require a subsidy, an entrepreneur would pay a premium for the right to build and operate a full land-grant railway under the cost and demand conditions

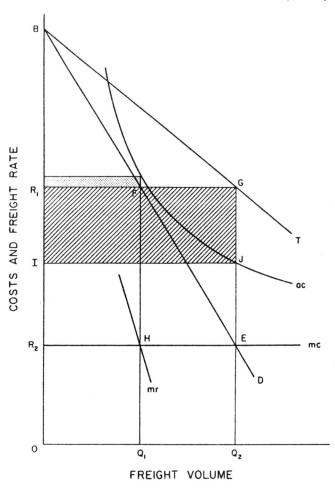

Fig. 3 Rail franchise value

of fig. 3. The maximum size of the premium is given by the hatched rectangle IR_1GJ and is explained presently. Meanwhile note that a subsidy by way of full land grant is excessive to the extent of that maximum premium. Selling the land-grant railway franchise to the highest bidder in an auction so as to capture the premium[12] was not done (at least, not openly — congressmen were alleged to have been bribed by budding railway entrepreneurs). Rather, a system of partial land grants was used, with effects significantly different from those of full grants.

Railways and land values

The size of the maximum premium to be paid for a full land-grant railway monopoly (hatched area in fig. 3) is derived as follows. With the freight rate set at R_2, operating costs equal freight receipts. An increase in freight rate above R_2 reduces land values by more than it increases railway operating profits; any decrease below R_2 causes a marginal operating loss in excess of the consequent increase in land values.[13] Therefore the best that could have been done by an integrated enterprise owning both railway and land would have been to set the rate R_2, and generate an increase in land rental value equal to the area of triangle $R_2 BE$. It is convenient to display this land value increase as a rectangle, so as to compare it with the fixed costs, K, by constructing the curve BT to which the demand curve BFD is marginal (so, in particular, area $OR_1 GQ_2$ = area $OBEQ_2$). When R_2 is charged, the excess of receipts — land rents plus freight — over all costs, including K, is thus the hatched area of fig. 3.

This premium also measures the net economic benefit ('social savings')[14] of the railway, and is, consequently, a guide for investment. A negative premium means the line generates more aggregate costs than benefits: average cost, ac, lies everywhere above BT. With a positive premium, all rail costs not met by freight receipts could be covered without resort to an 'external' subsidy from, say, taxpayers in general. An 'internal' subsidy, paid by shippers as renters or owners of land, can be arranged so that every customer-cum-landholder would benefit from operation of the line, with none being forced to pay more than what rail service is worth to him or her.

IV. Land-grant railways

This section discusses three aspects of American land grants: the question of excessiveness, their partial nature, and sale versus rent of land. In the case of fig. 3, a full land grant would give the rail entrepreneur more than the minimum necessary to achieve economic efficiency in investment and rating. Generally, American railways were granted only alternative blocks on either side of the line (along with other aids). Although a partial land grant reduced the subsidy, it had the disadvantage of discouraging some economically desirable lines.[15] Also, freight rates tended to be set above marginal rail cost. In fig. 4 the line BT_3, lying half-way between BT and the demand curve BD, shows the total freight plus land revenue for a 50 per cent land-grant railway. The curve BD_3 is marginal to BT_3, so the most profitable freight rate was now R_3. Compared with a full grant, land revenues fell by $R_2 R_3 UE$, of which the shaded triangle represents 'deadweight' economic loss, while the rectangle represents a transfer to rail operating profits from customers. The excess profits earned by the franchise, equal to the hatched rectangle, must be less than the corresponding profit from a full land grant ($IR_1 GJ$). Partial land grants were inefficient, but would, in general, have been excessive.[16]

There is an additional comment to make about private land grants, having to do

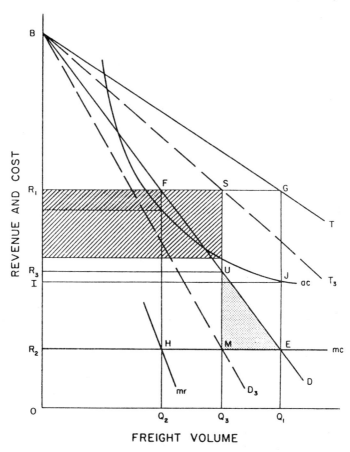

Fig. 4 Partial land grant

with the so-called problem of successive monopoly. Political pressures were put on American railways to sell their land rather than rent it out for long periods. Retention was criticised as 'speculation', and interpreted as an effort to force up the price. Interestingly, the lines could have had powerful incentives to sell rather than rent land, depending on the expectations of potential purchasers. And, once the land was sold, the railways then had a motive to alter freight rates, and dash those expectations. Assume that the line was maximising the annual combined value of *rent* on its grant and operating profit or loss[17] by charging the freight rate R_3. Immigrants and young men going west often preferred to own, not rent, the land they worked. *If purchasers of land assumed that freight rates would continue to be R_3, then* they would be willing to pay up to the capitalised value of R_3BU for the land. Of course, the rate R_3 could be sustained only if some of the land sale proceeds were 'earmarked' to cover future rail losses. Once the land was sold, however, the line had no incentive to charge anything but the ordinary monopoly freight rate, R_1. Alternatively, the original rail entrepreneurs, having first sold off the land to naïve purchasers, could then sell the line to another set of owners (who would have been willing to pay up to the capitalised value of the shaded rectangle for the right to run the rail monopoly even if an annual payment, K, is required to pay off a construction loan. Without any such debt, the annual monopoly profit from the line is R_2R_1FH. Fortunes were made by collapsing railway companies, selling off the assets, and starting again with a new legal rail entity). The new owners could show that, even if they had paid less for the line than it cost to build, they could not show a profit except at rates considerably above R_2: to charge only marginal cost would be to charge below full cost. In pushing up the rate the line would have imposed capital losses on landowners whose rate expectations were confounded. The advent of government rate regulatory commissions (as in the 1860s) could not solve the problem whenever the value of the land grants had been removed from the railway companies. Only if land were rented on relatively short-term leases, or sold with *binding agreements* as to freight charges, would the land grant system have given rise to the 'correct' economic incentives, those for marginal-cost rating of lines, and for the building of all those lines with positive economic value — and only those lines. What was needed was an 'earmarking' of the land values so as to enable the line to charge at marginal cost. Without such earmarking, and to the extent that land grants were sold off at prices that incorporated too optimistic expectations of future rail freight rates, the land-grant system encouraged the building of lines with negative economic value.[18]

V. Public ownership of land and railways in Australia[19]

Apart from a few early lines, the railways of Australia serving the rural landholder have been owned by the same governments that owned the land. This section explores why public ownership did not always produce efficient investment and rating, but ultimately tended to cause over-expansion.

We will distinguish two phases of operations, according to whether the systems were expected to cover costs from traffic receipts — as were the first lines built by governments, financed by London loans — or not. Initially each line was to be a commercial proposition. Quite early, however, common freight rates were set for all lines within a colony, so that users of profitable routes were subsidising others. Nonetheless the stated objective, largely attained, was that each colonial system cover costs without government budget subsidy and without credit for land value increases. Figure 5 illustrates the first phase, and shows how the cross-subsidisation, although not the best policy, may have increased economic efficiency while allowing an expansion of the rail system. The finances of two separate lines are shown, back-to-back. BD_1 is the demand curve for rail services in the first ('high demand') area, and BD_2 in the second. The railway lines are assumed to have the same, constant operating cost, shown as mc, and the same capital costs; ac is the common average cost curve. If each line covered costs from freight, only the first would have operated, with rate R and output Q_1. A common freight rate R' was struck, however, which produced profits in the high-demand area equal to the losses on the second line (these are shown as the two hatched areas); all costs were covered by freight receipts. In the case illustrated, the new combined output, $Q'_1 + Q'_2$, exceeded Q_1, the output before cross-subsidisation. Also, a second line that was economically justified was now constructed, albeit at the expense of rating the first line inefficiently (that is, above marginal cost). Overall, economic efficiency was improved, as fig. 5 has been constructed, because the hatched deadweight loss in line 1 (shaded triangle) is less than the excess of land values over costs from line 2 (also shaded; as before, the demand curve BD_2 is 'marginal' to BT_2). Neither the increase in aggregate output nor the improvement in economic efficiency was inevitable, although the former

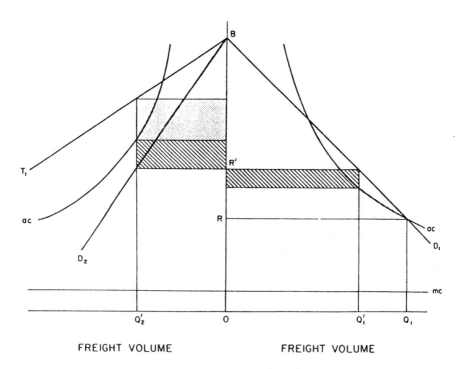

Fig. 5 Public railways covering costs

was likely, being of direct benefit to construction engineers and railway workers, both powerful political groups.

The restriction on railway construction implied by the constraint that costs be covered by freight receipts proved not to be effective in the face of pressure from those and other political interests. The first New South Wales railway commissioner proposed in the 1860s that his department should receive land grants; as in the United States; his proposal was rejected, as were later suggestions that land departments subsidise railways.[20] The extra land value, however, did accrue partly to the government treasury, as land rents and as taxes, at a time when the chief users of rail services (sheep pastoralists) dominated the parliamentary process. Pressure grew for further railway building, with the supporting arguments that the government could afford it because railways generated some kinds of what we now call 'external benefits' and, therefore, that new lines did not have to make a profit. By the end of the nineteenth century the argument in favour of new lines depended on the desirability of encouraging closer settlement, and the production of commodities other than wool (especially of wheat: new varieties, learning-by-farming, and decreases in ocean shipping costs made the export of wheat from New South Wales a possibility). Closer settlement and cultivation required the filling in of trunk-line rail systems with branch lines. In the commissions and parliamentary committees investigating each proposed branch line the practice grew up of comparing the expected annual loss after interest payments to the acreage of wheat (or other crops) that the line would bring forth, or to the value of that wheat (see section II above).

The combined effects of bringing land values into railway accounts and the continued use of state-wide freight rates are shown in fig. 6. To simplify the presentation, the two regions are assumed to have had the same demand curves (BD_1 and BD_2) and the same mc curves. Line 2 generated less economic benefit than full cost, regardless of how it was operated. Although line 1 was economically desirable, it could not cover costs without a subsidy being added to freight receipts. If all land value increases had been eaten up in a subsidy to its operation, at break-even its output would have been Q_1, far in excess of the optimum output, Q_0. When the finances of the two lines were combined, a common break-even freight rate, R', after land revenue attribution, produced outputs of Q'_1 and Q'_2. The surplus of land value over costs of line 1 was equal to the deficiency of the second line (shaded areas). In this extreme case, break-even after land revenues, output had expanded so far beyond the optima (zero for line 2, Q_0 for line 1) that all the potential excess of economic benefits over costs had been dissipated. A full land grant, in effect, when combined with a zero profit constraint, reduced net economic value of the system to zero.

Various experiments were tried in order to rationalise and restrict railway building and operations, including the levying of special taxes on local landholders to finance new lines, and the granting of permission to the railways to charge supernormal freight rates on some losing lines. Neither of these methods was widespread. A more

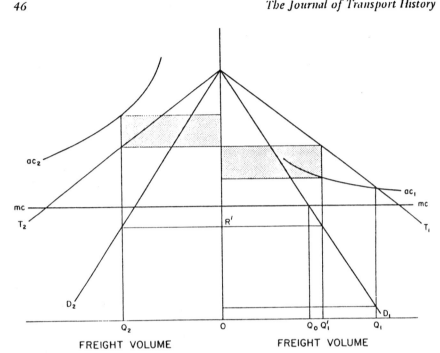

Fig. 6 Public railways dissipating land values

interesting technique originated in Victoria, as did many innovations in public administration. In the 1890s legislation was passed to allow the railway commissioners to claim a 'recoup' from the treasury for any losses imposed upon them by government directives.[21] No recoup claims were made for many years, despite the fact that the commissioners were constantly complaining that the government was requiring the building of new, unprofitable lines. The reason is simple: Parliament in those years always appropriated for the railways an amount sufficient to cover overall railway losses, whether they were caused by government initiative or not. There was no effective railway budget constraint.

Without such a budget constraint the railway system could have received a subsidy greater than the equivalent of a full land grant. Whether or not this was the case in more recent decades, it was probably not so by the late 1920s, when the building of new railway lines virtually ceased (until revived by the mineral discoveries of the 1960s). There were, in effect, some claimants for the economic profits that railways brought. Firstly, land rents were not always set at market rates, so that some portion of the increased land value stayed with the landholder. (Of course, landholders able to obtain a service without bearing the full cost lobbied for the building of rail facilities in their districts.) Next, other claims on government were successfully pressed, and, in particular, rural economic interests pushed for irrigation projects and, later, for motor roads. But the precedent of unprofitable railways was again followed, and these new 'development works' were not required to break even. The state became regarded, in the famous phrase, as 'one vast utility' for which only an overall, and long-run, budget balance was required.[22]

In addition to rural interests seeking public works, there were other claimants for fiscal benefits.[23] Relief from taxes was made possible by government budget ease. And, finally, the political power of the pastoral and ancillary interest diminished in the face of growing urbanisation and the rise of working-class political movements. This brings us to one last comment on the Australian case. In recent decades governments have begun to close down unprofitable lines, with the guiding principle that each service should recoup all costs through freight receipts: average, not marginal-cost, pricing is required. Thus the same kind of disappointment of farmers' expectations about freight rates as occurred in the United States is occurring in Australia. (Some private mining companies are also facing the violation of implicit or explicit freight rate contracts.)

VI. Conclusions

The simple model of railroad economics developed here, besides helping in the explanation of some characteristics of railway history in America and Australia, also points to a problem in public choice. This problem is how to internalise the marginal benefits generated by a utility like a railway in order that investment decision and pricing rules be economically efficient. Failure to 'solve' this public choice problem has consequences for the economic and political actions of various groups in society, especially the railways, and landowners and renters.

On first consideration, a land-grant, private railway system seems ideal, because it is in the interests of the entrepreneurs to price at marginal cost so as to maximise the combined value of the land and the line; moreover, no line would knowingly be built that did not promise an overall profit, that is, was not economically justified. A difficulty arises with the land-grant system, nevertheless, in that entrepreneurs who intend to sell rather than to rent land have an incentive to convince prospective land purchasers that freight rates will be low; once the land has been sold, however, the line then wants to break that promise, and to raise rates. Assuming this contractual difficulty is resolved, a full land grant generates incentives for efficiency, but creates no surplus for the 'rest of society' unless the land-grant franchise is sold.

Where the railway and the landlord are one, the efficiency problem appears solved: the greatest economic benefit would arise when there is a maximisation of the value of the surplus of land rents over railway losses, such as would be sought by a Leviathan-like government. In fact, where the railway and landlords are one government, there is a tendency for the 'profits' (including land rents) from economic lines to be used to subsidise losses made elsewhere. A government rail system can be vastly over-expanded, with the complete dissipation of the social savings from public enterprise prevented by the success of other claimants to the revenue generated by railways and other public works.

Notes

1 H. N. Scheiber, *Ohio Canal Era: a Case Study of Government and the Economy* (Athens, Ohio, 1969).

2 N. G. Butlin, *Investment in Australian Economic Development, 1861–1900* (Cambridge, 1964), p. 367, and see 'Monopoly lost', in N. G. Butlin, A. Barnard and J. J. Pincus, *Government and Capitalism: Public and Private Choice in Twentieth Century Australia* (Sydney, 1982).

3 E. A. Pratt, *The State Railway Muddle in Australia* (London, 1912); F. W. Eggleston, *State Socialism in Australia* (London, 1932).

4 S. L. Engerman, 'Some economic issues relating to railroad subsidies and the evaluation of land grants', *Journal of Economic History* (June 1972), 443–63.

5 I.e. the excess of benefits enjoyed by purchasers over the outlays made to attain those benefits.

6 We assume that a uniform plain surrounded the railway, and that rail freight could be loaded at any point. For products valuable in relation to their bulk or weight (e.g. whisky or wool) the catchment of rail extended beyond a day's return trip by waggon; we ignore this complication.

7 It is likely that a farm located far from the railhead would send fewer tons per acre than one near the railhead, because a distant farmer receives less for his product after rail and waggon charges than does a farmer close to the rail, and so the former concentrates more effort on activities that are not so transport-intensive. Such a decline in railed output per acre would tend to make the demand curve in fig. 2 more convex (viewed from the origin) than that in fig. 1. There are, however, other considerations. If each railhead were located on a uniform plain, the number of acres at any given distance from the railhead would increase with the square of that distance, making the demand curve in fig. 2 more concave than that in fig. 1. Now, this second effect disappears if produce can, costlessly, be loaded at any point along the rail line. The relationships, derived below, between land values, rail freight rates and waggon transport costs carry over when we translate PYD into a conventional demand curve.

8 In fig. 2 the increase in product net value is shown by area $Q_0 Q_1 IM$, and equals $P(Q_1 - Q_0)$. By inspection, we have $L = N > P[(Q_1 - Q_0) - V]/2$, where the shaded area V is equal to the rail charges on a quantity $Q_1 - Q_0$ at rates R_0. Of course the land rent increase must then be much less than half the market value of the increase product carried by rail.

Cf. E. F. Renshaw, 'A note on the measurement of the benefits from public investment in navigation projects', *American Economic Review* (September 1957), 652–63.

9 J. Dupuit, 'On the measurement of the utility of public works', *Annales des Ponts et Chaussées*, VIII (1844), in D. Munby (ed.), *Transport* (London, 1968), pp. 19–57; H. George, *The Land Question: What it Involves and How Alone it can be Settled* (New York, 1893).

10 The life of the line is assumed known and fixed; demand and cost conditions remain steady during that life. Also, rail entrepreneurs are assumed to purchase land at values that do not reflect at all the prospect of a railway; the pre-rail price is conveniently taken to be zero.

11 The shaded loss area, with $R_1 F$ as base, arises as the difference between the current account profit $(R_2 R_1 FH)$ and the annual fixed charge.

12 Cf. H. Demsetz, 'Why regulate utilities?', *Journal of Law and Economics* (1968), 56 ff.

13 Note that R_2 is the efficient freight rate in that the economic benefits, flowing from a rise above or decrease below R_2, fall short of the economic costs.

14 The measure of railway 'social savings' made by Fogel *et al.* corresponds to L in fig. 2. Calculations using fig. 2 show the social savings due to a fall in the long-distance freight rate from R_1 to R_0 is equal to the following: the increase in gross market value of farm output, less the increases in farm production costs, less the increase (or plus the decrease) in long-distance freight costs (equals earnings), less the increase in local cartage outlays.

15 For example, if the ac curve in fig. 4 cut BT but not BT_3, then a 50 per cent land grant would fail to elicit construction of a line that would be built with a 100 per cent land grant. More generally, too little investment may be made (including too slow a rate of construction). Note, however, that if some of the land were already owned by farmers, it would still be in their collective interest to subsidise the railway until it operated at freight rate R_2 (although, naturally enough, they would prefer to shift the subsidy burden on to others). Engerman (*loc. cit.*, 448–9) argued that, in the case of a natural monopoly, a subsidy should not be paid for the purpose of inducing investment. With this exception, the analysis of the present section can be regarded as a spelling-out of his discussion; see also J. B. Buchanan, 'A public choice approach to

public utility pricing', *Public Choice* (December 1968), 1–17; D. J. Brown and G. M. Heal, 'Two-part tariff, marginal cost pricing, and increasing returns in a general equilibrium framework', *Journal of Public Economics* (1980), 25–49; S. Damus, 'Two-part tariffs and optimal taxation: the case of railway rates', *American Economic Review* (March 1981), 65–79. The question of the optimal number of railway lines is left to a subsequent paper. In the present analysis no line existed that competed for freight from the same catchment.

16 II. Fleisig concludes that the grants to the Central Pacific 'were "excessive" at the margin, where excessive describes subsidization that influenced neither the decision to invest nor the speed of construction': 'The Central Pacific Railroad and the railroad land grant controversy', *Journal of Economic History* (September 1975), 553. For Fleisig's calculation of rates of return the subsidies may as well all have been lump-sum. The significance of land grants, however, is their incentive effects on investment and output (pricing) decisions. To repeat, the efficient arrangement is to auction the full land-grant franchise. Competition would then go as far as is possible to eliminate excessive profits. To the extent that prospective entrepreneurs did in fact bid for rail monopolies by bribes to congressmen, etc., competition would again tend to reduce profits to 'normal', even though the taxpayer in general does not benefit. Such bribes could not appear in Fleisig's calculation. The excessive profits that remained after bribery were captured by rail entrepreneurs and others, even enhanced, through financial manipulations (e.g. via construction companies, of which Credit Mobilier is the most famous; watering of stock; insider land trading; expropriation of private bondholders).

17 With the rate equal to marginal cost, an operating loss occurs if mc is decreasing. This complication is dealt with in J. J. Pincus, 'Railroads and Land Values', Discussion Paper, Center for Study of Public Choice, Virginia Polytechnic and State University, 1981.

18 This analysis suggests two reasons, possibly minor, why later land grants were sold off more slowly than earlier ones. Prospective purchasers may have bcome more aware of the threat of successive monopoly, and assumed that future freight rates would be closer to the ordinary monopoly level than to marginal rail freight cost. To this extent, the land grant was more valuable if rented, not sold. Also, as long as railways owned the land, rail entrepreneurs had less to fear from rate reductions, regulatory or otherwise, because these would raise their land rents and values.

19 This section draws upon the fuller discussion in Butlin, Barnard and Pincus, *loc. cit.* (1982).

20 For an analysis of the economic arguments of the time (including ones similar to those in this paper), see C. Goodwin, *Economic Enquiry in Australia* (Durham, N.C., 1966).

21 R. Wettenhall, 'The recoup concept in public enterprises', *Public Administration* (London, 1966), 391–413.

22 The history of New South Wales is interesting in this regard. By the 1890s it was virtually a free-trade colony, relying much more than Victoria on land for government revenues. With federation, New South Wales gained its share of federal import duties, the level of which depended on railway development; customs duties were like a form of land tax. New South Wales increased its railway network the fastest of all the states during the next two decades, and embarked upon very costly irrigation and other rural schemes.

23 An implication is that the US Homestead Act of 1862, by setting a maximum price to be obtained by government, may have reduced the likelihood of the subsidisation of uneconomic lines from sales of land located near economic lines.

Acknowledgements

This paper has benefited especially from the comments of Geoffrey Brennan, Joe Reid, Jr., Stefano Fenoaltea, and Paul David. The Centre of Policy Studies, Monash University, is to be thanked for financial assistance.

5 Private enterprise or public utility? Output, pricing and investment in English and Welsh railways, 1879–1914[1]

P. J. CAIN

I

It is not a matter for surprise that the heroic age of building before 1870 has always attracted the bulk of the attention of railway historians: but in the last twenty years there has been a revival of interest in the fate of the railway industry in late Victorian and Edwardian days. Professor Ashworth opened the modern debate in 1960 by investigating why it was that the profitability of the industry declined steadily after 1870.[2] This question has remained at the centre of the discussion ever since, and in the process a wealth of information on such matters as productivity, investment and pricing policy has been unearthed and evaluated.[3] The debate has, however, always been severely restricted by a lack of data on railway output. The companies produced figures for the tonnage of goods traffic originating on their system and the number of miles run by their trains but the output of a railway can only be calculated satisfactorily if the average distance travelled by goods traffic is known. Ton miles (tonnage carried × average length of haul) are the only reasonable measure of railway output and once they are available it is possible to compute that crucial measure of efficiency the train load (ton miles ÷ by the number of miles run by trains) and find out the price of railway goods services, the rate per ton per mile (gross goods receipts ÷ ton-mileage). Ton-mile statistics were not, in fact, systematically collected by the companies until 1920, with the single exception of the North Eastern, which compiled them for its own purposes in the Edwardian period and during the first world war. All the arguments about railway efficiency and declining profitability have therefore been conducted against the background of highly unsatisfactory statistics such as tonnage originating, train mileage and receipts and expenses per train mile.

Only one estimate of output after 1870 has been made, by G. R. Hawke in his now famous book *Railways and Economic Growth in England and Wales, 1840–70* (Oxford, 1970). Hawke worked out a ton-mileage series for the industry between 1870 and 1890 in the process of assessing the long-term fruitfulness of investment undertaken by the companies before 1870. This series does not appear to have been commented upon, or even mentioned, by anyone interested in the controversy over post-1870 railway growth and efficiency. The purpose here is to evaluate Hawke's figures critically and to use them as a starting point for some tentative suggestions about output, investment and pricing after 1870 which, it is hoped, might add a little to our knowledge of the performance of the railway industry.

II

The initial assumption made by Hawke, upon which his post-1870 series depends, is the rather fragile one that the average rate of haul on English and Welsh railways was thirty-three miles in 1865.[4] This implies an average train load of 55·7 tons and an average rate per ton per mile of 1·21d. Hawke then proceeds to estimate output after 1865 by various methods, using also the official figures for 1920. The resulting series depends crucially upon the assumption that train loads and average rates of haul increased steadily after 1865. Table 1 gives Hawke's ton-mile figures for certain years and spells out some of their implications.

One of the aims of the rest of this section is to suggest that these figures tend to overestimate the growth of railway output before 1890.[5] Leaving aside for the moment the implied steady rise in train loads, it is noticeable that Hawke's estimates indicate a rapid fall in the price of goods transport, a fall of a similar magnitude to that in prices generally:[6] yet this was the time when railway rates were supposed to be 'sticky' and when businessmen were complaining bitterly of a rising real cost of railway transport. The implied average rate of haul is also too high, and, in investigating this, it is possible to go further and make some suggestions about output growth after 1890, where Hawke's series ends. In order to do this we must work back, somewhat tortuously, from the official figures of 1920.

The average rate of haul on British railways in 1920 is given as 57.7 miles[7] but this is the result of adopting a tonnage figure arrived at by counting traffic conveyed over more than one railway (through traffic) once only. Before 1913 figures were collected on a tonnage-originating basis (i.e. through traffic was counted as a part of the tonnage of all railway companies over which it passed) and involved an element of double counting. Tonnage-originating statistics can, however, be obtained from the official railway returns of 1920, and it is also possible to disentangle the figures for England and Wales from those for Scotland.

The average haul of 37·9 miles for tonnage-originating traffic may be compared with Hawke's estimate of 40·5 miles for 1890 and it raises the obvious question: did the average haul fall after 1890? It did indeed do so, but from a rather lower level than Hawke suggests, and the figure of 37·9 miles for 1920 was reached only because of a substantial shift upwards in average hauls during the first world war.

The first piece of evidence which suggests that average rates of haul were substantially lower in 1913 than they were in 1920 is provided by the North Eastern. Its average haul rose from 24·8 miles in 1913 to 34·5 in 1920, an increase of 39·1

Table 1 *Hawke's output estimates, English and Welsh railways, 1871–90*

	1 Tonnage carried (million tons)	2 Train miles (millions)	3 Goods receipts (£ millions)	4 Ton miles (millions)	5 Implied train load (tons)	6 Implied rates per ton per mile (d)	7 Implied average haul (miles)
1871	140·4	74·0	22·4	4,700	63·5	1·14	33·5
1880	200·4	97·4	30·5	7,500	76·6	0·98	37·4
1890	259·1	122·3	36·0	10,500	85·9	0·82	40·5

Note. Col. 5 = col. 4 ÷ col. 2; col. 6 = col. 3 ÷ col. 4; col. 7 = col. 4 ÷ col. 1.
Sources. Cols. 1–3, *Railway Returns.* Col. 4, G. R. Hawke, *Railways and Economic Growth,* p. 92.

Table 2 *Tonnage originating and ton mileage, 1920*

	Tonnage (million tons)	Ton miles (million tons)	Average haul (miles)
Great Britain	511·8	19,173·0	37·5
England and Wales	448·3	16,975·4	37·9
Scotland	63·5	2,197·6	34·6

Source. Railway Returns for 1920. Tonnage-originating figures for the larger companies of England and Wales are given on pp. 264–5 and total 437·6 million tons or 60 per cent higher than the 273·6 million tons arrived at by avoiding double counting. By the latter method the remaining small companies of England and Wales can be shown to have carried 6·7 million tons, and the tonnage-originating figure has been estimated at 60 per cent higher, i.e. 10·7 million tons.

per cent.[8] If all railways increased their average hauls by two-fifths in wartime, then the figure for England and Wales in 1913 would have been about twenty-seven or twenty-eight miles (27·3 miles if the North Eastern's experience had been exactly reproduced elsewhere).

A similar figure for pre-war hauls can also be arrived at by a different method. During 1901 and 1902 George Paish, the editor of the *Statist* and a leading advocate of reform in railway accounting methods and management, made ton-mile estimates for a number of the larger companies. The results were published in the *Statist* and later in his book *The British Railway Position* (1902). Paish appears to have had a certain degree of inside information,[9] including access to preliminary ton-mileage figures for the North Eastern, that company having just decided to compile them on a regular basis. Besides this, Sir George Gibb, the North Eastern general manager, wrote a preface to the book and it is significant that, although he questioned a number of Paish's assertions about railway reform, he did not criticise the ton-mile estimates. Paish's data may be tabulated as in table 3.

Paish's five companies were responsible for approximately 55 per cent of goods tonnage originating on railways in England and Wales in 1900 and 65 per cent of the ton mileage of those railways in 1920. The average haul on these five companies in 1920 was 43·7 miles, compared with an overall figure of 37·9 miles. If we assume that the same relationship between these five companies and the rest held in 1900, then the average haul on all railways in England and Wales in 1900 would have been 26·4 miles. This is near the estimate of 27·3 miles for 1913 based on North Eastern data. The credibility of the 1900 figure is strengthened by the fact that the average haul on the North Eastern changed very little between the turn of the century and the war, rising from 22·2 miles in 1900 to 23·1 miles in 1911 and 24·1 miles in 1913.[10]

Table 3 *Ton-mile statistics, five leading English and Welsh companies, 1900–01*

	Ton miles (millions)	Train loads (tons)	Average haul (miles)	Rates per ton per mile (d)
LNWR	1,517·8	69·1	34·1	1·19
Midland	1,555·3	57·8	38·0	1·14
GWR	1,294·2	57·8	34·5	1·06
NER	1,171·5	66·6	22·2	1·24
LYR	444·8	71·0	20·7	1·57
All five companies	5,983·6	62·8	30·4	1·19

Source. G. Paish, *The British Railway Position* (1902), pp. 147–50, 172–6, 191, 195–7, 216–20, 281. Figures for the LNWR, GWR and Midland are based on 1900 freight and year ending June 1901 for receipts and train miles. NER figures are for 1900 throughout and for the LYR for the year ending June 1901 throughout.

The evidence accumulated so far indicates an average haul between 1900 and 1913 which was about 40 per cent less than that achieved in 1920. It is worth investigating why this was so, and the easiest way to begin is to look closely at the changes in North Eastern traffic before and after the war, the evidence for which is contained in appendix 1.

A study of appendix 1 shows that the effect of war was

1. To increase the long-haul goods traffic relative to the short-haul mineral traffic.
2. To increase the amount of longer-distance foreign traffic relative to local traffic and to increase the distance travelled by the foreign traffic.
3. To increase the average hauls even on the company's short-haul local traffic.

The reason for this change must be, in part, the reorganisation of railway working during the war, when the system was operated as far as possible as a unit. Equally important, however, was the fact that during the war a great deal of long-distance traffic which had hitherto gone to its destination by sea was sent by rail because of a catastrophic decline in coastal shipping facilities and colossal increases in shipment rates.[11] Moreover, much of this traffic taken over from coastal shippers had initially been lost to sea competition between the late 1880s and the turn of the century: average hauls before 1900 were nearer to the 1920 figure than they were between 1900 and 1913.

English railways were particularly susceptible to competition from coastal shippers. In such a small area none of the major centres of population was far from the sea. Four of the six largest conurbations in England and Wales — Greater London, south-east Lancashire, Merseyside and Tyneside — were actually centred on a major port, and these areas accounted for 30 per cent of the total population in 1911. Many other large towns and cities were also ports, for example Bristol, Cardiff, Swansea, Hull, Plymouth, Portsmouth and Southampton. Many major areas of population had, therefore, a direct connection by sea, and, besides this, the big inland towns and cities had the choice of sending their long-distance traffic either direct by rail or consigning it a short distance by rail to a port and sending it on by sea. One contemporary authority described the situation thus:

> ... this competition of the coasting services with the railways [was not] limited only to traffic between the ports directly concerned. Each and every one of these ports, great or small, might be and invariably was a collecting and distributing centre for coastwise traffic within a given radius or in connection with towns extending along the coast in either direction. In the case of certain ports, such as Goole and Gloucester, this inland distribution was facilitated by their connection with exceptionally efficient canal services. Then the railway rates from, say, London to some inland point not far from the coast for commodities which might otherwise go partly by water, were governed by the shipping freight from London to a convenient port plus the cost of road carriage by a local carrier.[12]

The competition of coastal shippers was then extended further by developing through facilities whereby the coastal shipper quoted overall rates which included the cost of short-haul rail traffic to and from ports concerned.

> With the combination of rail transport and coasting service, the original radius of, say, twenty miles when goods were conveyed by road from port to domicile was generally extended to one of at least fifty miles when they were collected or distributed by a supplementary rail journey: though in this respect the interests of the railway companies were not always identical since it would be to the advantage of, for example, a southern company to convey to a port on its own system traffic for Scotland, which if it went all the way by rail, that company might not handle at all, while the northern companies would be deprived of a long haul. It was generally the case of two short hauls by rail and a long sea journey being substituted for a throughout journey by rail.[13]

Pressures from sea competition were intensified by sharp falls in shipping freights after 1870. Tramp shipping rates (which include both ocean-going and coastal shipping traffic) moved as in table 4.

Table 4 *Tramp shipping rates (1869=100); quinquennial averages*

1871–75	106	1886–90	68	1901–05	51
1876–80	92	1891–95	58	1906–10	49
1881–85	74	1896–1900	64	1911–13	68

Source. B. R. Mitchell and P. Deane, *Abstract of British Historical Statistics* (Cambridge, 1962), p. 224.

Evidence of the sharp impact coastal shipping made upon railway long-haul traffic can be found by examining the changing structure of London's coal traffic after 1870.[14] The railway companies' share of this trade increased steadily from mid-century until 1876–80, when they accounted for 63·1 per cent of the total imported into the metropolis. Their share then declined marginally in the 1880s and was 61 per cent of the total in 1886–90. A much more precipitate decline in the railways' share was registered in the following decade. Figures for coal traffic into London by rail between 1892 and 1897 are unfortunately missing or unavailable but the downward trend is clear, for the railways accounted for only 48·6 per cent of the traffic in 1898 and 45·7 per cent in 1905. Their share increased in 1906–08 but then fell off again and was only 46.8 per cent in 1911. The companies not only lost out relatively to coastal shippers; the absolute amounts carried to London by rail also declined from a peak of 8·2 million tons in 1892 to 6·95 million in 1898. Only in one year thereafter (1907) did the tonnage of coal carried into the London market by rail exceed the 1892 figure. The decline of the traffic was facilitated by the abolition of London port dues in 1890 but it also reflects the intensity of competition as well as changing market structures and, it appears, rising railway rates. A leading coal-trade newspaper summed up the position in 1910 in this way:

> The reason for the growth of the seaborne coal trade as compared with that carried on by the railways is to be found in the rise of railway rates which effectively keeps the large output from the North Midlands out of the London market. For some years, the South Yorkshire gas coal collieries did a considerable trade with the metropolitan gas companies through the medium of a special train load rate granted to them by the railways, but when this was discontinued and the ordinary rates re-established the collieries found it impossible to retain the business and it reverted to Durham... Moreover, the increasing use of electricity and gas for power and heating purposes, both by manufacturers and householders, must operate rather in favour of those districts whose natural avenue to London is by water...[15]

This loss of position in the London market from the late 1880s was extremely significant for the fate of long-haul traffic in particular. Coal was the single most important commodity carried by the companies and represented upwards of two-fifths of all goods tonnage at this time; and London, as by far the biggest centre of population and remote from the coalfields, was the centre of the outstanding long-haul traffic in coal. It is surely reasonable to suggest that the railways lost a considerable amount of long-haul traffic in other low-value bulk commodities, especially minerals, to London and that they also lost a significant part of their long-haul traffic to other places, especially East Anglia, the South and the South-east, which were abundantly supplied with good port and harbour facilities.[16] The loss of a considerable portion of the long-distance traffic would not only have reduced overall hauls significantly but also put upward pressure on average rates by lowering the proportion of the most cheaply rated traffic.

Taking the changing fortunes of rail long-haul traffic in relation to sea competition as a crucial determinant of changes in average hauls, it appears that hauls were rising, albeit slowly, until about 1880, that they eased slightly in the next decade, fell quite sharply after 1890 until 1900 and then rose slightly.[17] (The continuing decline which might have been expected after 1900 as coastal shipping competition intensified was probably offset by a cessation of a certain amount of inter-company competition and an increase in through traffic as a result.) With this in mind, we can suggest an (extremely tentative) picture of changes in railway output after 1870, working backwards from a twenty-eight-mile estimate for just before the first world war (table 5).

Table 5 *Output estimates, English and Welsh railways 1871–1911*

	Tons carried (millions)	Train miles (millions)	Goods receipts (£ millions)	Ton miles (millions)	Train load (tons)	Rates per ton per mile (d)	Average haul (miles)
1871	140·4	74·0	22·4	4,212	56·9	1·28	30
1880	200·4	97·9	30·5	6,212	63·5	1·18	31
1890	259·1	122·3	36·0	7,773	63·6	1·11	30
1900	359·5	153·3	45·3	9,707	63·3	1·12	27
1911	448·3	130·9	52·3	12,552	95·9	1·00	28

There is no need to labour the point that, at best, this table offers some rough guide to trends rather than gives a precise outline of changes in the railways' fortunes. What it suggests is rapidly rising output and train loads in the 1870s accompanied by falling rates; and a slower growth of output, a tendency for train loads to decline and rates to fall in the 1880s, followed by stagnation in both train loads and rates after 1890, the latter reflecting the loss of much low-rated bulk long-haul traffic in that decade.[18] After 1900 output rose no more rapidly than between 1880 and 1900; but the very sharp increase in train loading indicated in the table does confirm the well known fact that 1900–13 was an age of railway reform.[19] The decline in rates after 1900 also fits in with the evidence available from other sources.[20] In the rest of this paper some of the implications of these trends need to be investigated.

III

We may begin with prices and pricing policy, starting with Hawke's model, which was formulated with the pre-1880 period in mind. Hawke begins with the premiss that the companies were discriminating monopolists, able to charge different prices for the same services to different groups of consumers. He also makes 'the usual assumptions about public utilities, namely, a large capital investment and low, approximately constant (marginal) variable costs', which follow from the fact that

Private enterprise or public utility?

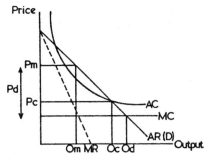

Fig. 1 G. R. Hawke, *Railways and Economic Growth* (1970), p. 351

'the plant is not used to physical capacity'.[21] He then produces (p. 351) a diagram (fig. 1).

Output under simple monopoly would be O_m and price P_m; under competition O_c would be produced in price P_c. But the discriminating monopolist can produce output up to O_d at price P_d, that is, the prices charged could be anywhere between P_m and P_d, the highest price being the limit of monopoly power, the lower floor of pricing being set by the need to cover the marginal costs of new traffic. Discriminating monopolists can, therefore, increase output beyond that attainable under competition or simple monopoly. Hawke argues that the railways did just this before 1880 as, for example, when they lowered rates in order to bring distant suppliers into a market hitherto dominated by the geographically favoured.[22]

He is careful to restrict his arguments to the period before 1880 and evidently believes that after that date the tendency for rates to fall was arrested to some degree.[23] His model is nonetheless only a more rigorous and elegant version of the argument, dear to railway economists after 1870, that the railways charged 'what the traffic will bear'. Acworth, the chief theoretician of the railways in the late nineteenth and early twentieth century, always argued that, since most of their costs were fixed, it would pay them to carry anything which covered their very low variable costs: any revenue above that level would make some contribution to the fixed expenses. The matter of primary importance to railways, therefore, was to 'get traffic' in the belief that 'any rate ... is better than no rate'.[24]

Acworth's position — which reflects to a large extent what railway managers said in public that they were doing[25] — suggests that the companies continued to push rates down after 1880 in order to increase traffic which they assumed automatically would be profitable. Railway rates, however, do not seem to have fallen as rapidly as one would expect if this were the case, and Acworth himself offered one qualification of his general position which is of interest in this connection. In 1897 he wrote that:

> Long distance through rates have fallen largely and will continue to fall because the potential increase of traffic may make any low rate ere long more profitable than a high one. Local rates

remain unaltered. They are not reduced because it is considered that all the existing traffic does pass at existing rates and there is no potential traffic to be developed by reduction.[26]

Now the long-distance traffic to which Acworth refers was that most subject to coastal shipping competition after 1880 and it was this competition which was the principal reason for the pushing down of rates. When faced with claims of undue preference and unfair discrimination the companies argued strongly that the long-distance traffic was cheaper to handle but they were also clear that the main reason for lowering rates was pressure from the sea.[27] Moreover they do not seem to have pushed their rates down far enough, since, in the 1890s in particular, they lost a considerable part of their through traffic business; and, as rates were lower on long-distance traffic than on short, the overall effect of sharply falling hauls was stagnation in the average rate per ton per mile in the 1890s.

After 1880 the companies were also trying hard to maintain their local rates and might well have raised many of them, as Acworth admitted, had it not been for the pressure of commercial opinion.[28] Until the 1870s the erection of the network and the diversion of traffic from other agencies meant that the companies could indeed lower rates in the expectation of stimulating demand greatly. After that time, with the network largely complete, they found that traffic grew steadily and was less price-elastic; at the same time they were faced with a new competitive force, coastal shipping, which took a large slice of their most profitable long-haul traffic. The growth of the economy after 1870 was also slower than in mid-century and the companies had to accommodate themselves to a fragmented, short-haul and relatively high-cost freight, a reflection of the small-scale nature of British industry as a whole. In the coal trade, for example, the companies had to cater for a multiplicity of small collieries and even smaller dealers — who stubbornly resisted their frequent attempts to rationalise the traffic[29] — at the same time as they were losing a good deal of their most lucrative runs to London and the south coast. The 'retail' character of much of British railway traffic was commented on frequently by contemporaries,[30] and, in consequence of it, the companies came to see themselves as providers of a high-cost service upon which rates could not be cut without disaster. This determination to maintain local rates or even to raise them ran counter to the expectations of their customers. The sharp dichotomy between the huge railway company and its small-scale business clients made the former seem particularly threatening and reinforced the latter's intense hostility to 'monopoly'. As profits were squeezed after 1875, arguments about the need to control railways because of this monopoly power hardened into a demand that, as agencies supposedly immune from normal market forces, they should act as public servants.[31] The companies, as creatures of parliamentary statute, were peculiarly vulnerable to this pressure from trading interests, the more so as the 'railway interest' in the Commons was slowly losing its power.

The companies' response to these pressures to reduce local rates was to maintain them as far as possible but, at the same time, to increase the range, and frequency, of their services. Hence the rapidly growing train mileage which was a feature of railways after 1880, and the provision of a host of ancillary facilities such as free wagon use, free or below-cost warehouse services, generous settlements in the face of claims for negligence or damage and so forth. Basically the increase in services meant that, in lieu of lowering rates, the companies were willing to absorb some of the costs of small businesses, raising their own in the process. It was at this point that the companies began to lose sight of the importance of good train loading,[32] something to which table 5 bears emphatic testimony.

Essentially, then, the companies responded to public pressure for rate reductions by improving their services and increasing their costs (raising the MC line in Hawke's diagram). Average costs would also have been increased by the relative decline in long-haul traffic from the late 1880s. The intention, therefore, was not so much to stimulate output by increasing ton mileage as to increase train mileage relative to ton mileage. The policy had the advantage of offering the companies an outlet for their competitive urges other than the — possibly self-destructive — method of price competition; and it also reduced the occasions on which they could be accused of undue preference, something which must have increased had rates been more flexible. Disguised subsidies offered as part of service completion offered less opportunity for controversy than did more open forms of price cutting.

No doubt the companies adopted service competition rather than competitive rate cutting because they felt that it did less harm to their prospects of profit. In this they may well have been mistaken. The policy appears to have been based on the Acworth-type assumption that the basic condition of the industry was excess capacity and that additions to the service could be absorbed without involving much additional cost. This was true of the companies' rural mileage, where increases in train mileage did not put any strain upon existing lines and installations. It was markedly untrue of railway operation in the region of major towns and cities, which accounted for the bulk of the traffic carried.[33] Here the steady growth of new traffic together with increased services to placate the public was constantly leading to demands for new track, widening, additional warehouse space and station accommodation, and so forth.

What is missing from the Acworth approach — and perhaps from the calculations of managers — is a clear distinction of this kind with its implications for costs and prices. In areas where the companies were working near to or at capacity, additions to train mileage could easily involve heavy capital expenditure to avoid congestion. In these cases charges should, ideally, have been higher and related to long-run marginal cost and its investment implications.[34] Given public hostility to railways this was impracticable even if railway managers had perceived the problem — which is doubtful. The policy of increasing services ahead of traffic led to a steady rise in capital expenditure and was the principal reason for the fall in the ratio of gross revenue to paid-up capital in Great Britain from 9·25 in 1870–74 to 8·56 and 8·78 in 1895–99.[35]

The mediocrity of this performance can best be gauged by contrasting the English

Table 6 *English and German capital expenditure and output (£ millions)*

Year	Paid-up capital	% increase	Gross revenue	% increase	Working expenditure	% increase	Net revenue	% increase
English and Welsh railways								
1882	633·8		56·6		29·7		26·9	
1906	1,043·1	+64·6	100·3	+77·2	63·1	+112·5	37·2	+38·3
Prussian State Railways								
1882	219·0		25·1		13·8		11·3	
1906	475·0	+116·8	93·4	+272·1	58·5	+323·9	34·9	+208·8

Sources. For English railways, *Railway Returns*, and G. R. Hawke and M. C. Reed, 'Railway capital in the United Kingdom in the nineteenth century', *Economic History Review*, 2nd ser., XXII (1969), p. 272; for Prussian railways see C. H. Pearson and N. S. Reyntiens, 'Report on Railways in Germany', which is appendix IV of *Railway Conference Reports* of 1909, P.P., 1909, LVII (Cd 4677), p. 68. Marks have been converted to pounds on the basis of 20 marks to £1.

performance with that of the Prussian railway system. The comparison is by no means exact — it is not easy to be sure that the item 'total capital' is the same as 'paid-up capital' for example — but the figures do give a strong impression of the sharp difference in the productivity of capital on the railway systems in two different industrial countries.

The most obvious inference to be drawn from table 6 is that while increased capital expenditure generated very rapid increases in gross revenue in Germany it did not do so in Britain. While the ratio of total revenue to total capital rose from 11·5 to 19·7 between 1882 and 1906 on the Prussian railways, the gross revenue/capital expenditure ratio in England and Wales moved from only 8·93 to 9·62 in the same period. The difference reflects the dynamic nature of the German economy and the relative 'newness' of the railway system.[36] English railways, faced with slower traffic growth, would have had to restrict their capital expenditure severely, increase their rates, or curtail their services, in order to maintain or increase profitability, and as we have seen this was hardly feasible politically at a time when the companies were coming to be regarded as public services. Under pressure from the traders, they warded off decreases in rates by offering a more frequent and costly service and paid for it in low train loads, unprofitable capital expenditure and falling profitability. The latter remained possible without painful consequences so long as the returns on equivalent investments, such as government stocks, remained low enough to ensure that the companies could still have recourse to the stock exchange for investment funds.

If service competition was the most important reason for increasing capital expenditure it was not the only one. The capital account was increased on occasion in order to satisfy managerial needs for power, glory and perquisites. The building of high-cost urban stations in mid-century is an obvious example;[37] the extension of the Manchester Sheffield & Lincolnshire's line to London in the 1890s is probably another. Nor should one underestimate the 'non-economic' reasons for continuing to reduce overall profitability, by building new lines, for example. Railways, like motor cars in the twentieth century, became a necessity even for those who could not afford them. Local squires and landowners felt sometimes that they could maintain or buttress their local authority and prestige by linking their area with the wider world, motives which appear to have played a part in the building of the Golden Valley Railway and some of the lines in the remoter parts of Scotland.[38] In other cases, lines promoted by businessmen were built not in the expectation of direct profit (except perhaps through speculative selling of shares) but as a means of increasing the growth of local trade in general. The promotion of the Hull & Barnsley Railway in the 1880s was undertaken by businessmen who believed that the North Eastern's policies were reducing the profitability of business in that part of south Yorkshire.[39] The pressure of public demands and the fear of competition also pushed large companies either to extend lines into poorly peopled areas or to absorb their unprofitable small neighbours.

In the history of railway enterprise the divide between policy in late Victorian times and the Edwardian period is sharp. There was a dramatic fall in dividends at the turn of the century, accompanied by shareholder revolts, an inability to raise capital at acceptable rates of interest and a rising trend in prices of railway raw materials. This shock coincided with the recognition by the companies that the Railway and Canal Traffic Act of 1894 would make it more or less impossible to raise rates in the future. After 1900 the companies were intent on restoring their profit margins by the only method left open to them, that of lowering their costs. In this, as Irving has shown, and as the tentative train-load figures in table 5 indicate, they had remarkable success. What the reform meant, in essence, was that they managed to increase ton mileage significantly without the help of a rise in average hauls and with a significant decrease in train mileage. This was accomplished through a combination of internal reforms and a diminution of inter-company service competition.[40] As a result, the ratio of gross revenue to paid-up capital rose from 8·78 in 1895–99 to 9·67 between 1910–12, the latter being the highest figure ever recorded.[41] The increase in profits was not great, however, since working expenditure was pushed up by steadily rising prices. It should also be noted that the traders, who were losing many of the hidden subsidies of 1880–1900, were appeased by the fact that after 1895 prices were rising, rates edged downwards and therefore the real cost of transport was falling. There is a sense in which rigid rates at a time of rising prices were the price the companies themselves had to pay for the reorganisation of their services.

By the time the first world war broke out, though, they were near to exhausting the gains they could make from reforms of railway working, given the existing structure of the industry and prevailing attitudes about the public responsibilities of railway companies. Further reform required, *inter alia*, further consolidation and amalgamation (achieved in 1921) and greater charging flexibility, something only achieved in very recent times. Alfred Marshall, the economist, reviewing the state of the industry in 1919, wrote:

> a fairly old railway, holding the greater part of the transport of a compact industrial district, is likely to have so completely adjusted its appliances to the traffic, that each of them is well occupied; and it does its work so economically that any addition to that work should have to carry nearly full costs. In such cases cost of service could automatically become the chief regulator of railway charges.[42]

'What the traffic could bear' was a reasonable charging maximum for sparsely populated areas but not for congested urban ones. Marshall went on to argue that, on this criterion, many railway rates did not contribute their full share to railway expenses and that, for example, facilities were sometimes reduced or badly limited in areas where the companies had a complete monopoly to compensate for this. More important, he believed a low-rate policy might 'encourage the expansion of industry and trade in unsuitable places'[43] and that some railway traffic, including some bulk, low-value traffic, should ideally have gone by sea or even via a revived canal system.[44]

Marshall's analysis was reasonable in theory but impossible in practice. He underestimated the fund of hostility to a more flexible charging policy;[45] and his whole analysis presupposed a degree of planning and integration between transport services which has been out of the question right up to the present day.

IV

The arguments about railway efficiency after 1870 have revolved around three different approaches. Aldcroft has seen the poor profit performance of the companies principally as a result of unenterprising management. Pollins, in his turn, plays down the idea of any profit crisis until the late 1890s, whilst he attributes the difficulties which arose then to rising costs and wages rather than to inefficiency on the part of management; and he was also quick to point out that productivity increased rapidly after 1900. The third approach, that of Ashworth, Irving and Gourvish, is markedly different in that, although inclined to the view that there was a serious long-term problem of profitability, these historians all point to pressures beyond the control of the railways which forced them into unprofitable ways. As Gourvish recently put the matter, recent research has thrown 'the debate about falling profitability back on to the constraints imposed by parliament, traders and the public, and the "public service" image the railways were increasingly forced to adopt, rather than on to managerial shortsightedness and inefficiency'.[46] The conclusion of this present article must be that this latter position is, broadly speaking, correct.

Appendix 1 *North Eastern Railway, ton miles and average hauls*

	1913			1920		
	Ton miles (millions)	% share	Average haul (miles)	Ton miles (millions)	% share	Average haul (miles)
Mineral traffic						
Local traffic	702·1	73·5	16·2	592·1	57·4	18·8
Foreign forwarded	70·7	7·4	47·2	384·2	37·3	69·4
Foreign received	181·7	19·0	44·6	54·5	5·3	38·0
Through traffic	1·0	0·1	16·3	0·3	–	12·2
All foreign	253·4	26·5	45·0	439·0	42·6	62·8
Total	955·5	–	19·6	1,031·1	–	26·7
Goods and livestock						
Local traffic	283·4	42·6	23·6	227·3	27·0	24·8
Foreign forwarded	203·3	30·5	53·3	330·6	32·2	69·7
Foreign received	142·6	21·4	60·1	296·0	28·8	71·5
Through traffic	36·6	5·5	148·0	123·3	12·0	119·1
All foreign	382·5	57·4	59·4	762·9	73·0	75·6
Total	665·9	–	36·1	1,027·1	–	48·7

Source. BTHR, NER 4/134, *Traffic Statistics 1902–1925*, pp. 7–10, 13–14. Free-hauled traffic is included in the figures. The basis on which the ton-mile figures were compiled was slightly changed in 1920. On the 1920 basis the average haul on goods traffic would have been 39·4 miles in 1913 and for minerals it would have been slightly less in that year at 19·2 miles. The overall average haul would have been 24·8 miles in 1913 on the same basis instead of the reported 24·1 miles (*ibid.*, pp. 210–11). The changes do not affect the argument herein presented.

Appendix 2 Coal traffic to London by railway and coastal shipping, 1870–1911

Year	1 By rail (million tons)	% share	2 By sea (million tons)	% share	3 Total London coal traffic (million tons)	4 Coal shipments received coastwise (million tons)	London's share (%)
1870	3·76	55·7	2·99	44·3	6·75	7·28	41·1
1871	4·45	61·7	2·76	38·3	7·21	6·84	40·4
1872	5·00	58·7	3·55	41·5	6·55	6·55	54·2
1873	5·15	65·9	2·67	34·1	7·81	6·70	39·9
1874	4·69	63·2	2·73	36·8	7·42	6·53	41·8
1875	5·07	61·8	3·14	38·2	8·20	7·06	44·5
1876	5·17	61·2	3·27	38·3	8·45	7·10	46·1
1877	5·42	63·1	3·17	36·9	8·59	7·15	44·3
1878	5·59	63·6	3·20	36·4	8·79	7·12	44·9
1879	6·55	65·1	3·51	34·9	10·06	7·78	45·1
1880	6·20	62·5	3·72	37·5	9·91	7·74	48·1
1881	6·75	63·9	3·81	36·1	10·56	8·04	47·4
1882	6·55	63·1	3·83	36·9	10·38	7·91	48·4
1883	7·09	63·5	4·08	36·5	11·17	8·39	48·6
1884	6·85	61·5	4·29	38·5	11·14	8·45	50·8
1885	7·08	60·8	4·56	39·2	11·65	8·79	51·9
1886	7·13	60·4	4·67	39·6	11·80	8·81	53·0
1887	7·32	60·8	4·73	39·2	12·05	9·09	52·0
1888	7·61	60·9	4·89	39·1	12·52	9·39	53·0
1889	7·87	62·3	4·77	37·7	12·65	9·43	51·0
1890	8·16	60·8	5·26	39·2	13·42	9·78	53·8
1891	8·06	58·8	5·64	41·2	13·70	10·18	55·4
1892	n.a.		5·76		n.a.	10·38	56·6
1893	n.a.		6·36		n.a.	11·84	53·7
1894	n.a.		6·81		n.a.	11·97	56·9
1895	n.a.		6·86		n.a.	11·72	58·5
1896	n.a.		7·26		n.a.	12·45	58·3
1897	n.a.		7·50		n.a.	12·94	57·9
1898	6·95	48·6	7·38	51·2	14·29	12·38	59·6
1899	7·04	48·3	7·54	51·7	14·58	12·76	59·0
1900	7·74	49·2	8·00	50·8	15·74	13·05	61·3
1901	7·40	49·1	7·66	50·9	15·06	12·66	60·5
1902	7·36	47·7	8·08	52·3	15·44	13·64	59·3
1903	7·10	47·1	7·97	52·9	15·07	13·58	58·8
1904	7·14	46·3	8·29	53·7	15·43	14·29	58·1

	1		2		3	4	
Year	By rail (million tons)	% share	By sea (million tons)	% share	Total London coal traffic (million tons)	Coal shipments received coastwise (million tons)	London's share (%)
1905	7·14	45·7	8·49	54·3	15·63	14·55	58·4
1906	7·60	47·6	8·73	52·4	16·33	14·40	58·2
1907	8·35	50·5	8·20	49·5	16·55	14·17	57·9
1908	8·19	49·5	8·21	50·5	16·40	12·94	57·7
1909	7·81	46·8	8·90	53·2	16·71	14·92	59·6
1910	7·71	46·3	8·98	53·7	16·69	15·33	58·7
1911	8·02	46·8	9·17	53·2	17·19	15·62	58·8

Sources

Col. 1. (a) 1870–89: *Mineral Statistics*, published by the Geological Survey until 1881 and in Parliamentary Papers thereafter. They are reprinted up till 1879 in B. R. Mitchell and P. Deane, *Abstract of British Historical Statistics* (Cambridge, 1962), p. 113. (b) 1890 and 1891: these are estimates compiled from monthly and quarterly data by the coal-trade newspaper the *Colliery Guardian*. See vol. 61, 16 January 1891, p. 111, and vol. 63, 15 January 1892, p. 112. (c) 1898–1911: from 1898 to 1905 figures are available from the *Minutes of Evidence and Appendices of the Royal Commission on Canals and Waterways*, P.P., 1908 (Cd 3718), appendix 24; and *ibid.*, 1909 (Cd 4840), appendix 3. Figures for 1901–09 are printed in the *General Reports on Railways Capital, Traffic, Working Expenditure, etc.* published annually in Parliamentary Papers; and between 1903 and 1911 figures are available in the Greater London Council's *London Statistics*, XIX–XXII (1908–12). There are one or two slight discrepancies in the figures where sources overlap. The *General Reports* give 7·16 million tons for 1905 while the other two sources give the total printed in the table above; and in *London Statistics* the 1906 tonnage is given as 7·57 million.

Cols. 2 and 4. 1870–1911: *Mineral Statistics* until 1897 and *Statistics relating to Mines or Quarries* thereafter. The figures for sea traffic into London are also given for 1898–1911 in the *R.C. on Canals and Waterways* and *London Statistics*. Col. 3 excludes canal traffic (only 34,000 tons in 1911) and col. 4 is for England and Wales only. The figure for 1876 in col. 4 is an estimate.

A graph of coal traffic into London by sea and by rail from 1837 to 1911 is given in P. Bagwell, *The Transport Revolution* (London, 1975), p. 73. This is not based on exactly the same sources as the table here presented. It omits the figure for sea traffic into London for 1897 and has no figures for coal traffic into London for 1890 and 1891. The explanatory comments on the graph are also somewhat confusing.

Notes

1 I should like to thank the participant of the 1980 Midland Economic Historians' Conference held in Nottingham and the anonymous referee who read the article for helping me to improve both the argument and the presentation.
2 W. Ashworth, *An Economic History of England, 1870–1939* (1960), pp. 120–6. See also the same author's 'The late Victorian economy', *Economica*, XXXIII (1966).
3 The chief works here are: D. H. Aldcroft, 'The efficiency and enterprise of British railways, 1870–1914', *Explorations in Entrepreneurial History*, V (1968); H. Pollins, *British Railways: an Industrial History* (1971), ch. 5; R. J. Irving, *The North Eastern Railway Company 1870–1914: an Economic History* (Leicester, 1976) esp. ch. 12; R. J. Irving, 'The efficiency and enterprise of British railways, 1870–1914: an alternative hypothesis', *Economic History Review*, 2nd series, XXXI (1978); T. R. Gourvish, 'The performance of British railway management after 1860: the railways of Watkins and Forbes', *Business History*, XX (1978).
4 Hawke, *Railways and Economic Growth*, p. 62. He also offers a much more solidly based estimate of thirty-four miles for the average rate of haul on the coal traffic in the same year. *Ibid.*, pp. 182–6.
5 Hawke does suggest that, if his figures for 1870–90 are correct, and given the figure for 1920, the rate of growth of output between 1890 and 1920 must have been slower than previously. *Ibid.*, p. 62.
6 The Sauerbeck and Rousseaux price indices indicate falls in prices between 1871 and 1890 of 28 per cent and 24 per cent respectively, compared with Hawke's implied fall in rates of 28 per cent. The price indices are in B. R. Mitchell and P. Deane, *Abstract of British Historical Statistics* (Cambridge, 1962), pp. 472, 474.
7 P.P., 1920 (Cmd 1430), p. 265.
8 B.T.H.R., NER 4/134. *Traffic Statistics 1902–25*, p. 14, gives average haul in 1913 as 24·09 miles but, adjusted to a 1920 basis, the figure is 24·8 miles. *Ibid.*, pp. 210–11.
9 Paish, *op. cit.*, p. 16.
10 Average haul figures for the North Eastern are reprinted in Irving's *The North Eastern Railway Company*, p. 235, and in P. J. Cain, 'The British railway rates problem, 1894–1913', *Business History*, XX (1978), p. 90.
11 E. A. Pratt, *British Railways and the Great War* (1920), pp. 272–4; D. H. Aldcroft, 'The eclipse of British coastal shipping, 1913–21', *Journal of Transport History*, VI, 1 (1963). Tramp shipping rates rose from 68 in 1913 to a peak of 751 in 1918 and were still at 374 in 1920 (1869 = 100). Mitchell and Deane, *op. cit.*, p. 224.
12 Pratt, *British Railways*, pp. 265–6.
13 *Ibid.*, p. 266. The italics are mine.
14 The table on which the figures in the following paragraph are based is printed in appendix 2.

15 *Colliery Guardian*, 99, 6 May 1910, p. 871. A thorough analysis of railway coal traffic would have to consider not only changing demands for different kinds of coal but also shifts in the geography of supply.
16 The loss of part of the London market would have been the most crucial, however, since the share of the metropolis in coal traffic received coastwise in England and Wales rose from 51·9 per cent in 1885 to 61·3 per cent in 1900 before falling off slightly. These figures are derived from the table in appendix 2.
17 It is interesting to note that Paish's estimates of ton mileage for the London & North Western also indicate a falling average rate of haul between 1880 and 1900, though only of 4·7 per cent. Paish, *The British Railway Position*, p. 28.
18 The figures presented in table 5 suggests a rate of growth of output of 4·4 per cent 1871–80, 2·3 per cent 1880–90, 2·2 per cent 1890–1900 and 2·3 per cent 1900–11. Hawke's figures imply 5·3 per cent 1871–80 and 3·4 per cent 1880–90.
19 Receipts per train mile on UK railways rose by 37 per cent between 1900 and 1911, compared with an implied rise of 52 per cent in the train load figure in table 5 and an increase of 90 per cent in train loads on the North Eastern. Irving, 'The profitability and performance of British railways, 1870–1914', p. 66, and his *The North Eastern Railway Company*, p. 241. See also Table 3.
20 On this see Cain, *op. cit.*, pp. 89–92. The implied fall in rates between 1900 and 1911 in table 5 is about 10·7 per cent, compared with a fall of 6·8 per cent in the figures for the North Eastern.
21 Hawke, *Railways and Economic Growth*, p. 350.
22 *Ibid.*, p. 334.
23 *Ibid.*, p. 91 and p. 353 n. 1.
24 W. M. Acworth, 'Railways in their economic and financial aspects', *Journal of the Institute of Bankers*, XVIII (1897), p. 2. Quoted in Irving, *The North Eastern Railway Company*, p. 130 n. 73. See also Acworth's *Elements of Railway Economics* (Oxford, 1905), pp. 71–6.
25 See, for example, Sir G. Findlay, *The Working and Management of an English Railway* (1894), p. 264.
26 W. M. Acworth, 'The theory of railway rates', *Economic Journal*, VII (1897), p. 329.
27 P. J. Cain, 'Railways and price discrimination: the case of agriculture, 1880–1914', *Business History*, XVIII (1976).
28 Acworth, 'The theory of railway rates', p. 329.
29 See, for example, the evidence given to the *Royal Commission on Coal Supplies*, P.P. 1904, XXIII (Cd 1991), qq. 11,700, 12,024, 12,124, 12,126, 12,128, 15,374.
30 See E. A. Pratt, *Railways and their Rates* (1905), pp. 91–7.
31 It has been argued that, well before 1900, the companies acted 'for the most part as if they were public utilities'. D. E. C. Eversley, 'The Great Western Railway and the Swindon Works in the Great Depres-

sion', in M. C. Reed (ed.), *Railways in the Victorian Economy: Studies in Finance and Economic Growth*, (Newton Abbot, 1969), p. 134.

32 It should also be noted that the small traders had some influence in keeping up average rates in that pressure from them made it impossible for the companies to offer lower rates for train loads. J. Mavor, 'The English railway rates question', *Quarterly Journal of Economics*, VII (1893–94), pp. 309–11. Irving, 'The profitability and performance of British railways, 1870–1914', p. 55.

33 *Departmental Committee on Railway Rates and Preferential Treatment*, P.P. 1906, LV (Cd 2959), pp. 8–9.

34 For a modern look at the problem see K. M. Gwilliam and P. J. Mackie, *Economics and Transport Policy* (1975), esp. pp. 81–3, 102–4.

35 Irving, 'The profitability and performance of British railways, 1870–1914', p. 48.

36 It is noticeable also that the ratio of gross operating revenue to net capitalisation on railways in the United States rose from 13·7 in 1890 to 19·2 in 1912: *Historical Statistics of the United States*. Comparisons between British railways and those of other industrial countries are overdue.

37 As illustrated, for example, in G. Channon, 'A nineteenth-century investment decision: the Midland Railway's London extension', *Economic History Review*, 2nd series, XXV (1972).

38 C. L. Mowat, *The Golden Valley Railway* (1964); N. T. Sinclair, 'The Aviemore line: railway politics in the Highlands, 1882–1898', *Transport History*, II (1969).

39 Irving, *The North Eastern Railway Company*, pp. 118–20.

40 P. J. Cain, 'Railway combination and government, 1900–14', *Economic History Review*, 2nd series, XXV (1972), p. 629.

41 Irving, 'The profitability and performance of British railways', p. 48. Conscious efforts to improve train loading appear to have begun in the mid-1890s, if the London & North Western is typical. See C. E. Grasemann's evidence to the *Dept. Committee on Railway Accounts and Statistics*, P.P. 1910, LV (Cd 5052), esp. para. 77 and p. 488. See also B.T.H.R., LNW 4/62, *Mean Average Annual Loading of Wagons, 1896–1905. Liverpool District*, which shows a 13·5 per cent increase in wagon loads 1896–1900 and a 10 per cent improvement between 1901 and 1905. The Board of Trade, commenting on railway performance in 1901, stated that 'it is probable that there has been in recent years some tendency towards a gradual increase in the average weight of a train-load', P.P. 1901, LXVII (Cd 749), p. 7.

42 A. Marshall, *Industry and Trade* (1919), p. 465.

43 *Ibid.*, pp. 474–5, 476.

44 *Ibid.*, pp. 476, 497–505.

45 As was demonstrated by the ferocious opposition to the 4 per cent general rate increase of 1913. See Cain, 'The British railway rates problem', p. 96.

46 Gourvish, 'The performance of British railway management', pp. 198–9.

6 The capitalisation of Britain's railways, 1830–1914

R. J. IRVING

I

This article re-examines the positions which economic historians have taken recently on the timing, extent and reasons for the heavy capitalisation of home railways in an attempt to resolve the inconsistencies in argument and interpretation that exist currently, and to settle also some unresolved issues concerning the business judgement of the industry's managers after 1870.

As a starting point for the analysis of the capital base of the industry, the article reworks some of the data on railway capital formation presented by Hawke and Reed[1] and provides a revised series of 'capital raised' for the period 1870 to 1890.

II

In 1969 G. R. Hawke and M. C. Reed published a series of the paid-up capital of United Kingdom railways from 1825 to 1912.[2] The article traced the course of railway capitalisation, and examined the geographical dispersion of funds and the methods employed to raise money. The analysis confirmed the accepted picture of railways making greatest demands on the capital market in the upward movement of trade cycles and spending funds behind the peak of the cycles. Generally, the capital market was shown to be efficient, directing funds to the most profitable opportunities irrespective of location. The paper identified also the historically high level of 'low risk' capital used to fund development and noted, in particular, the importance of loans and debentures in the early stages of the industry and in periods of financial disorder. In the later years of the century significant structural and regional trends emerged. These involved a gradual diminution in the proportion of funds raised through ordinary shares, but regionally there was a greater use of preference shares in Scotland and an increased use of loans and debentures in England and Wales.

Subsequently, however, it has been suggested that the data presented by Hawke and Reed may not portray accurately the amount of capital raised, because of a failure to identify and exclude nominal additions which companies made before 1890 to their capital accounts,[3] and that this omission may vitiate some of the authors' conclusions on the impact of railway development on regional capital markets and the proportions of funds raised on different classes of stock. Before any debate on the capitalisation of home railways can take place, these issues must be considered.

The source of the difficulty with Hawke and Reed's series lies in the extent to which the published statistics of paid-up capital diverge from a true measure of funds collected through the capital market. The Board of Trade's *Railway Returns*, used heavily by the authors in their analysis, only began to identify and separate out nominal additions to paid-up capital from 1890, at which time there was a rising trend of submissions to Parliament by railways desirous of manipulating large amounts of their capital through duplication and division of stock,[4] and some concern about the motives for these moves. Hawke and Reed's judgement was that it was only from around 1890 that the published series of paid-up capital diverged significantly from a true measure of funds collected, and they chose not to examine the position before that date. In fact, nominal additions were common long before 1890, even though the object of such additions was generally different from those which caused the Board of Trade in 1890 to begin quantifying the situation and demanding public examination of the motives of interested parties.[5] Analysis of the railway companies' own ledgers and accounts shows that significant nominal adjustments to paid-up values were made from an early stage. The records of the Taff Vale Railway, for example, show that as early as 1843–44 nominal additions of £25 per share were made to the holdings of 3,000 'old' £100 shares in the company in respect of capitalised interest,[6] whilst those of the London & North Western Railway record nominal adjustments to capital of almost £1 million in 1846 as part of the fusion of the three companies creating the new organisation.[7] It is clear, therefore, that pressures of early railway financing could create legally sanctioned nominal additions to capital and that the process of railway amalgamation also led to nominal additions as the reorganised undertakings sought to consolidate in revised, marketable securities the numerous stocks and shares of their constituent companies.[8] The restructuring of railway capital after 1890, therefore, emerges merely as a particular stage in a process which goes back to the 1840s and requires more careful examination than hitherto if a truer measure of capital collected is to be devised.

Quantification is possible without recourse to the accounts of every one of the eleven hundred or so companies that existed from time to time before 1900, since in 1890 the Board of Trade published its own attempt to calculate the amounts by which the capital of United Kingdom railways had been nominally increased up to and including that year.[9] Although the dates at which each company made changes to its capital are not recorded in the return, the document provides the basis for examination of the accounts concerned to establish this information; the cumulative results are shown for each company in appendices 1 and 2 and summarised below in aggregate for the years up to and including 1889.

Table 1 *Distribution of nominal additions to British railway capital to 1889 (£)*

	England and Wales	Scotland
Ordinary	8,890,605	12,067,106
Preference and guaranteed	21,856,183	7,029,066
Loans and debentures	7,661,977	50,301
Total	38,408,765	19,146,473

Source. Appendices 1 and 2 below.

By 1889 some 6·6 per cent of British railway capital was already 'watered'. Even more significantly, almost 17 per cent of Scottish railway capital had by this date been manipulated. The appendices show that, while the process goes back to the 1840s, major reconstructions were undertaken in both England and Scotland from around 1870, in particular, and that these are significant enough both in total and on a regional basis to justify a recalculation of Hawke and Reed's estimates before 1890 and a rejection of the view that it was only from 1890 that the published series of paid-up capital diverge from a true measure of funds collected through the capital market. Revised estimates of real capital collected in England and Wales, and Scotland, before 1890 are represented in appendices 3 and 4.

The most obvious feature of the nominal additions described in appendices 1 and 2 is that the real growth of railway capital between the mid-1870s and 1890 was rather more modest than previously assumed and the real rate of return on capital higher. The figures in table 2 show clearly that the gross return on capital was rising steadily decade by decade after 1870.

Since this has significance for the view that late Victorian railway managers were profligate with shareholders' funds we shall return to the point. As far as the structure of the capital account is concerned, we see that in England and Wales reconstructions of capital were concentrated disproportionately on preference and guaranteed stocks, some £21·9 million of these being watered by 1889 compared with £7·7 million of loans and debentures and £8·9 million of ordinary securities. Apart from the fact of the existence of dilution at this rate before 1890, an implication of the concentration on preferential securities is that the contribution of loans and debentures to real capital formation at this time has been understated. The graphical representation in Hawke and Reed's analysis of a sharp decline in the importance of

Table 2 *Gross return on capital raised: railways of Great Britain, 1850–1910 (nominal additions excluded)*

Year	Gross return (%)
1850	5·58
1860	8·31
1870	8·57
1880	9·39
1890	9·68
1900	10·81
1910	11·28

Note. The gross return is derived by expressing traffic receipts as a percentage of capital raised minus nominal additions.
Source. Hawke and Reed, *op. cit.*; B. R. Mitchell and P. Deane, *Abstract of British Historical Statistics* (Cambridge, 1962), pp. 226–7; appendices 1 and 2 below.

loans and debentures between 1870 and 1890 is somewhat misleading. The figures in appendix 3 suggest that between those dates the share of real funds raised through loans and debentures was pretty constant at 26–27 per cent.

The spate of reconstructions of preference and guaranteed stocks before 1890 came about as a result of a need, in the majority of cases, to bring different classes of these stocks to a uniform and lower rate of interest as the price of money fell in the later 1870s. Such operations took place on the London & North Western, Midland, Great Northern, Lancashire & Yorkshire, Great Eastern, Furness and London & South Western railways. In a minority of cases, involving struggling companies like the Cambrian, East London and Devon & Somerset railways, nominal additions were sanctioned by Parliament as parts of schemes of arrangement designed to reorganise the capital account and capitalise arrears of interest. Only towards 1890 do conversions in England and Wales begin to appear substantially on ordinary and debenture capital. That of 1889 involving the ordinary capital of the Taff Vale Railway attracted interest in that it represented a doubling almost of the paid-up capital of the company and was cited as an example of the reconstruction of capital for reasons to do more with the external value of the company's securities.[10] In the 1890s

The capitalisation of Britain's railways

further conversions of this sort were to arise in England involving, for example, the London & South Western and Great Northern railways. Overall, however, in England and Wales before 1889, nominal additions were not normally unusually large in relation to the total capitals of the companies involved and were generally concerned with internal management of the fixed-interest capital account rather than with external manipulation of price and the interests of investment trusts, stock jobbers and the like.[11]

In Scotland a quite different situation clearly existed and deserves comment. Firstly, it needs re-emphasising that Scottish railway capital was particularly subject to manipulation, with some 16–17 per cent 'watered' as early as 1889 and 26 per cent by 1912, compared with 6 per cent and 13 per cent for England and Wales. Here the elimination of nominal additions from the published estimates alters significantly the picture painted by Hawke and Reed of the level of capital formation in Scotland's railways in the mid-1870s and 1890s. Hawke and Reed identify a large jump in paid-up capital in Scotland in 1876 and on more than one occasion refer to Scottish railway capital growing at around 5 per cent annually in the mid- and later 1890s. In fact at both times the jumps are 'paper' ones only: in real terms we find almost zero growth in 1895–96 and low rates thereafter, thus lengthening significantly the time span of minimal growth in the industry's capital base. Between 1895 and 1905, for example, the growth of paid-up capital was approximately 1·6 per cent per annum and, over the period 1895–1912, 1·2 per cent. The concentration of nominal additions on share capital, especially the ordinary class, is in marked contrast to England and Wales and requires more comment than it has received. As in England, the effect of adjusting the raw data back to 1870 is to increase the contribution to funding of railway investment by loans and debentures which in terms of nominal adjustments were all but untouched in Scotland before 1890. This is important, since it suggests that the Scottish bias towards preference capital was a little less pronounced and the interest in loans and debentures rather greater than suggested. In the 1880s the real contribution of this class of money to Scottish railway development was consistently between 24 and 25 per cent, compared with the average of 21 per cent offered by Hawke and Reed. The overall concentration of nominal additions on ordinary capital in Scotland before 1890, of course, confirms the dependence of Scottish railways on fixed-interest securities of all kinds for fresh capital in the late nineteenth century, and the picture that emerges is rather depressing. Levels of real capital formation were much lower after 1870 than previously indicated, whilst there was also a heavier dependence on low-risk funds and not least a smaller share of total British railway capital. For 1880 the Scottish share of British railway capital excluding nominal additions was 12·3 per cent rather than the 13·3 per cent implied in Hawke and Reed.

Also present was a difficulty in managing the capital account. The evidence for this is not just the relatively high level of 'water' in the accounts from an early date, but also the character of the 'water' itself. From 1876 reconstructions of Scottish railway capital involved not just alterations to facilitate amalgamations and consolidation of different stocks, but also the contentious duplication of ordinary stocks which, long before it occurred in England, brought with it worries about the speculative manipulation of funds.[12] The prominent role by 1890 of the two leading Scottish companies in the stock conversion movement is reflective of the condition of the Scottish capital account and ultimately of the low profitability of the industry.

III

The aggregate statistics of capital formation having been modified, it is possible to proceed to a more accurate analysis of the capitalisation of home railways. In this respect, it has never seriously been disputed that the British railway system suffered from and was damaged financially by the burden of over-capitalisation. Railway analysts recognised the fact before 1914, identifying it as a cause of the disappointingly low return on invested capital and seeing in its existence a force depressing the level of railway pay.[13]

Explicit in most early analyses of the industry's performance was the assertion that home railways were over-capitalised from their inception almost, owing to the heavy cost of (1) the preliminary expenses of promoting and securing Acts of Parliament, (2) the cost of acquiring and holding land, and (3) excessive expenditure on the system. This last was seen to arise partly from the effects of competition and undue pursuit of engineering perfection, also from excessively rigorous engineering and directional regulations imposed upon promoters by Parliament and from building ahead of the real demands of traffic.[14]

This interpretation, the significance of which is its stress on the early origins of over-capitalisation, has a respectable pedigree, stemming largely from the great body of evidence placed before the several major parliamentary committees during the 1830s and 1840s.[15] These established a view of the chronology, causes, results and extent of excess capital in home railways which held substantially good in the parliamentary, financial, business and academic communities right down to 1909, when the Board of Trade Railway Conference concluded in one area of its deliberations that much of the problem was attributable to the outlay in obtaining statutory powers and in the acquisition of land.[16] Notwithstanding the considerable degree of unanimity in nineteenth-century opinion on the matter, and indeed of the similar views of contributors to the subject in the inter-war years,[17] the position of the post-war generation of British transport historians as a whole has been inconsistent and remains unsettled.

In 1952 Harold Pollins made the first examination of the railway companies' books to consider the Victorian argument that opposition to railways, the difficulty of obtaining Acts of Parliament and the high prices of land adversely affected capital accounts to a significant degree.[18] He satisfied himself that they did not, since as regards preliminary expenses the analysis showed that the great majority of companies spent less than 5 per cent of their total capital outlay on this item, allowing the conclusion that the financial results of obtaining Acts were not important and could be dismissed as an annoying interlude in railway financial history. As regards land, although the figure, at 13·9 per cent, was higher, the evidence indicated that it was swollen not so much by high prices as by the need of the companies to purchase rather more land than had been estimated. Excesses of expenditure over estimates were due rather more to technical difficulties in constructing the system and arose therefore on the more legitimate side of the business.

From this analysis a reinterpretation of the Victorian view began to emerge. Railways did not, it seemed, pay excessive prices for land. Preliminary expenses had been exaggerated. Major elements of the Victorian view were seemingly misplaced, and interest moved forward to the behaviour of railway companies in the later nineteenth century. Thus Ashworth noted in 1960 a 'great growth' of paid-up capital per route mile, from £40,000 in 1864 to £66,000 in 1904. 'It will readily be seen,' he wrote, 'that much of the high capitalisation of English railways derives not from reckless waste in promoting and establishing railways before 1850, which is popularly supposed to be the fount of most of their subsequent ills, but from the enlargement of traffic capacity in the last quarter of the nineteenth century'.[19] This very explicit statement was supported by Aldcroft who in two books[20] and an article[21] published in 1968—69 consistently argued that railways became overcapitalised in the late nineteenth century.

Around the same time, Pollins himself quoted with approval Ashworth's emphasis on the growth of capital after 1870 as the cause of the high capitalisation of the industry,[22] with M. R. Robbins also lining up in the same camp.[23]

Not all accepted the change of emphasis, though. In 1959 Savage remained squarely alongside the traditional position,[24] whilst in 1971 Bagwell inclined to the view that more money was spent in promoting the system than necessary and that this had long-run significance for costs.[25] In 1972, in a study of the London & North Western Railway, Gourvish drew attention to the pressing and enduring problems caused by the cost of railway promotion and construction and in 1980 briefly but unequivocally stated that high capital costs in the industry were established at an early stage.[26] Perhaps most important of all, in his examination of the impact of railway building upon Victorian cities and the land market Kellett argued in 1969 that the Pollins-initiated revisionism, though an interesting corrective to wilder allegations of landowners' rapacity, was nevertheless to some degree misplaced because the analysis actually had its sums wrong and understated the quantity of capital consumed in land purchase by the railways.[27] I myself have also argued from a different standpoint to Kellett's that the statistical evidence does not point to railways becoming over-capitalised after 1870.[28] It must also be pointed out that demonstrating that expenditure on land and preliminary expenses was not as great in relation to total capital as was once thought, or, in the case of land, was not due to high prices extorted by greedy landowners, does not prove that these items of expenditure were unimportant in increasing the capital burden on railways to a critical level.

On paper, then, the situation is confused. The purpose of this section of the article is to resolve the arguments. To anticipate, it will be reaffirmed that a high level of capitalisation was inherent in the system from the start and suggested also that the costs of preliminary expenses and land were, as nineteenth-century analysts insisted, significant contributors to the process. It will not be denied that there was some growth of paid-up capital per mile after 1870, but it will be shown that the amount was greatly exaggerated by Ashworth and the character of the investment misunderstood by others. The conclusions drawn also allow late nineteenth-century railway managers as a group to be relieved of the serious criticism that they spent their shareholders' funds unwisely and confirms the argument advanced elsewhere[29] that the main cause of diminishing returns after 1870 was inappropriate operating policies determined partly by the industry's political problems.

IV

In commencing the review of evidence it is necessary to remind ourselves of the chronology of capital growth. Using our modified statistics of paid-up capital, this is shown in tables 3, 4 and 5.

Looking first at the actual level of investment in each route mile of railway, we see from table 3 that by 1840 just under £32,000 of capital had been raised to fund

Table 3 *Capital and mileage, British railways, 1830–1910*

Year	Gross capital raised (£'000)	Route miles of way open	Nominal additions recorded (£'000)	Capital per mile, excluding nominal additions (£'000)
1830	1,825	97	–	18,814
1835	7,268	338	–	21,502
1840	47,570	1498	–	31,755
1850	228,580	6084	1,024	37,402
1860	327,736	9446	1,801	34,505
1870	502,634	13562	2,305	36,892
1880	694,527	15563	25,716	42,974
1890	850,047	17281	57,024	45,889
1900	1120,766	18680	186,473	50,015
1910	1256,151	19986	196,908	52,999

Source. Hawke and Reed, *op. cit.*; Mitchell and Deane, *op. cit.*, pp. 225–6; appendices 1 and 2 below.

the construction of each of the 1,500 miles of way open. By 1850 the figure had increased to £37,400. From 1850 to 1870 there was no great change as mileage increased to 13,500. After 1870 there is a growth of capital per route mile to £46,000 by 1890, to £50,000 in 1900 and to £53,000 in 1910. Allowing for nominal additions, each route mile carried some 35 per cent more capital in 1900 as compared with 1870. The first point to be made about these figures is that they roughly halve the extent of Ashworth's growth. Exclusion of nominal additions produces an altogether different result.

Confirmation of this point is found if we take an alternative approach, that of relating capital to *track* mileage rather than *route* mileage. Track mileage statistics were not returned systematically before 1903. Mitchell has argued,[30] however, that up to 1877 we can assume the ratio of track miles to route miles was 2:1. After 1877 the ratio increases at a rate of 0·1 per annum, the rate of increase shown in the published returns from 1903–12. This produces the figures in table 4.

Table 4 *Capital raised per track mile, British railways, 1850–1910*

Year	Capital raised net of nominal additions (£'000)	Track miles	Capital per track mile
1850	227,556	12,168	18,701
1870	500,329	27,124	18,445
1880	668,811	31,593	21,169
1910	1059,243	49,709	21,308

By this method, between 1870 and 1910 capital per mile increases by 15 per cent only, nearly all the growth taking place in the early 1870s. Significant though these figures are, they acquire more substantial meaning when they are related to railway revenue and used to give some indication of the earning power of the assets. Following Lardner and Clapham,[31] this is achieved by the expedient of expressing gross railway revenue as a percentage of paid-up capital, as in table 5, which extends the data presented in table 2.

Table 5 shows that from 1850 to 1910 the gross return on British railway capital was generally below 10 per cent and never as much as 12 per cent. This, of course, is before the cost of operating the system is deducted to yield the amount available for distribution or retention as working capital. Nevertheless, whereas the figures of capital per route and track mile show growth after 1870, the figures of gross return indicate that the earning power of assets was consistently higher after 1870 than before.

The British return may be compared tentatively with that achieved overseas. For America, often used as a comparator of better practice, indications are that between

Table 5 Gross return on the railway capital of Great Britain, 1843–1910

Year	Traffic receipts (£ million)	Gross capital raised (£'000)	Nominal additions recorded (£'000)	Gross return net of nominal additions (%)
1834	4·5	66,492	–	6·8
1850	12·7	228,580	1,024	5·6
1860	27·1	327,736	1,801	8·3
1870	42·9	502,634	2,305	8·6
1880	62·8	694,527	25,716	9·4
1890	76·8	850,047	57,024	9·7
1900	101·0	1120,766	186,473	10·8
1910	119·5	1256,151	196,908	11·3

Source. Hawke and Reed, *op. cit.*; Mitchell and Deane, *op. cit.*; appendices 1 and 2 below.

1850 and 1860 gross receipts were roughly 11 per cent of capital, around 1900 approximately 13 per cent and by 1912 of the order of 19 per cent.[32] In India between 1870 and 1914 the return increased from 7·7 per cent (1870/5) to average consistently over 10 per cent in the 1880s and 11·5 per cent between 1905 and 1910.[33] In Prussia the return on total capital seems to have been similar to that in America, at around 11·5 per cent in 1882 and over 19 per cent in 1906.[34] The gross return on British capital appears at the lower end of the performance ranges, therefore, but reflects the general trend for returns to improve over time.

What we see, then, are British railways burdened with relatively unremunerative capital which really only begins to fructify in any significant way for investors after 1860. However, the figures for home railways clearly question the position of those who have taken the view that the industry became over-capitalised after 1870. Importantly, the statistical evidence of gross return accords with the literary evidence to be found in the minutes of evidence of the parliamentary inquiries that examined railway affairs throughout the nineteenth century.

V

If we examine the subject matter of the major Victorian parliamentary inquiries into railways, it is clear that the question of the capital cost of constructing the system and the possible level of excess capital in the industry exercised the official mind early in the railway age rather than later.

The first inquiry really to review railway development was the select committee of 1839 on railway communication. It spent much time examining the extent to which original cost estimates had been exceeded and considering why. By 1844, when Gladstone's select committee sat, while there was a significant interest in competition and amalgamation, the interest in constructional costs was undiminished and indeed had widened its scope to embrace comparison with constructional costs in Europe and America. In 1846 the Committee on Railway Acts Enactments demanded that every railway make a comprehensive return of all aspects of its business. At the head of the list of questions to be answered was a demand for the length of main line and the cost of constructing same, to be returned under detailed standardised headings embracing preliminary, constructional and equipment outlays. Discussions with witnesses dwelt lengthily upon the comparative cost of construction of home and foreign railways. Repeatedly in the 1840s and 1850s committees were set up to consider how the allegedly extravagant and burdensome parliamentary expenses attendant upon obtaining Acts of incorporation could be diminished, the wider debate involving discussion of the contribution that the statutory procedures required in Britain were making for relatively high constructional costs. In the 1850s, too, Cardwell's committee of inquiry found it worth while to spend time debating whether the extent of excess capital in home railways was at that time £70 millions, as alleged by Samuel Laing (equivalent to not far short of 30 per cent of total capital) or £40 million to £50 million, as suggested by Robert Stephenson (equivalent to nearer 20 to 25 per cent of capital).[35]

By the 1860s the emphasis had shifted. The Royal Commission of 1865-66 showed little interest in the matter, concerning itself with pricing policy and amalgamation, and nobody disputed the passing assertion of the general manager of the South Eastern Railway that 'everybody knows the railways in England have cost more than twice the money the French and Belgian state railways cost' or that the difference was attributable 'to the cost of land chiefly and also to parliamentary expenses'.[36] The heavy capitalisation of home railways both absolutely and relatively was by this time an accepted fact. As early as 1850 Robert Stephenson had commented on the relative aspect in noting that 'the general characteristics of the English works are solidity and strength, durability and grandeur; of the Belgian, Prussian and others of the Continental works, simplicity, judicious economy . . . and of the American works rigid and parsimonious economy'.[37] Thus in the early 1870s parliamentary interest was dominated by amalgamation. In the later part of the decade concern was with accidents. In the 1880s an obsession with rates and fares gathered pace and was reflected in parliamentary business, as was the subject of the hours of work of railway labour. The growth of paid-up capital at this time attracted little attention. Although this was less true of the 1890s as diminishing net returns set in, whenever the question of the excess capital inherent in the system was raised, as in the Board of Trade conference of 1908, the principal cause of the excess was always likely to be seen as parliamentary costs and the continuing expense of acquiring and holding land.[38]

The available statistical and literary evidence, then, points to the early railway age as critical for capital costs and moreover points to the issues of preliminary expenses and land charges. It is to this period and these issues that we turn.

VI

In analysing total constructional costs, Pollins used two main sources. First there was the major parliamentary return of expenditure incurred by 186 railway companies up to the end of 1857. This yielded the conclusion that parliamentary expenses as a proportion of total outlay were in most cases not more than 5 per cent. Second, detailed statistics derived from the accounts of individual railways showed that the aggregate expenditure of twenty-seven selected companies on land totalled 13·9 per cent of total expenditure on construction, excluding the cost of locomotives and stations.

As regards parliamentary expenditure, the estimate of 5 per cent is confirmed by statistics published recently by Simmons which suggest that 5·6 per cent of the total outlay of ninety-three English and Welsh companies by 1849/50 had gone on this item.[39] As regards land, Pollins's method of calculating the share of land costs has been severely criticised by Kellett, who, using a parliamentary return of 1849/50, showed that land costs amounted to 16·5 per cent of all costs incurred in promoting, constructing and equipping the system and 25 per cent of construction costs. Simmons's figures, more comprehensive but from the same source, confirm Kellett's view and show that land costs for a larger sample were 17·2 per cent of all costs and 26 per cent of construction costs.[40]

Table 6 *Financial returns from ninety-three railway companies in England and Wales, 1849—50*

Cost of construction	£89,173,000
Cost of land	£23,254,000
Parliamentary and legal expenses	£7,565,000
Engineers' charges	£2,908,000
Plant	£12,315,000
Total	£135,215,000
Land as a percentage of all costs	17·2
Land as a percentage of construction costs	26·1

Source. J. Simmons, *The Railway in England and Wales, 1830—1914* (Leicester, 1978), pp. 43—4.

These returns are important, altering our perception of the situation. At the end of the day they still support a different but significant point made by Pollins, which is that the major element of constructional costs was the physical expenditure on the system itself, in engineering and equipping it. For Pollins this was the real significance of his analysis. Parliamentary expenditure to him was modest, and on balance the cost of land seemed to derive from a need to buy more than originally envisaged rather than price. Undoubtedly there is truth in this. The Great Western told the 1839 Committee that twelve acres of land per mile had been required rather than the eight estimated. The London & Croydon Railway confirmed it had no problem over the price of land but effectively ended up building a different railway to that estimated for. The same is true of the Newcastle & Carlisle line.[41] But while Pollins's conclusions are reasonable when taken on their own in relation to the proportion of parliamentary to total expenditure, and consistent with the evidence of a number of important witnesses to parliamentary committees in respect of land requirements, they nevertheless miss a fundamental point identified at some length by Samuel Laing of the Board of Trade in a detailed official submission to the select committee of 1844.[42] The point is that, compared with railway building elsewhere in the world, the whole institutional environment of railway promotion in Britain was seriously burdensome to the industry and contributed significantly to the relatively high cost of the system.

As noted, the official mind in the early railway age was concerned by the gathering evidence that the cost of constructing home railways was high. In 1839 the concern related mainly to the substantial divergence from initial estimates which formed the basis of parliamentary approval for construction. By the mid-1840s the concern was more broadly based and derived from a growing recognition that foreigners were building their lines much more cheaply and a worry that the outcome might be a high-cost system in Britain in terms of rates and fares and poor returns to investors.[43] The average cost per route mile in Britain, the Board of Trade told the 1844 committee, at £34,360 was 'eight times the average cost of the railways of the United States, three or four times the average cost of those of Germany, double the average cost of those of Belgium and one-half higher than the average cost of the most expensive railways in France'.[44] It set out to establish the causes of this situation.

First, under the heading of preliminary expenses, it was established that in Britain parliamentary expenses averaged approximately £700 per mile, the average concealing substantial fluctuations. The London & Brighton, for example, spent £3,000 per mile, the London & Birmingham £650. This particular item had no parallel on French or Belgian railways. Preliminary expenses under the heading of legal charges, engineering, direction and other incidental outlays in preparing legislation, at £1,600 per mile, did have a direct parallel overseas and emerged at £1,000 per mile more in England than Belgium. Roughly half the difference was attributable to the greater expense of legal proceedings in Britain, and to higher rates of professional emolument. The other half was due to the procedural systems governing railway promotion.

As far as land and compensation were concerned, the select committee noted that on home railways the average was £5,000 per mile. In France and Belgium it was £2,300 per mile. Of the difference, a portion was due to the lower real value of land on the Continent. A part, however, estimated at £1,500 per mile, reflected the expenses incurred by many undertakings in order to obtain support or buy off opposition and the excessive sums paid by way of compensation. 'A great deal of this excess,' it was noted, 'might probably under a different system have been avoided.'[45]

The estimate and reference to the systemic cause of the excess are important. They are not so much an attack on the rapacity of landowners as a comment on the institutional arrangements under which railway companies sought powers to acquire and hold land, and on society's attitude to railway promotion in relation to the safeguarding of property.

In Britain the railway network emerged in an unplanned manner by which authority for construction could be obtained only after protracted negotiations in Parliament. A key feature of this process was the investigation of each scheme to ensure that it complied with standing orders governing the proposed disposition of the line and its mode of construction. The standing orders, which were designed to protect the property interests affected by the proposed railway, facilitated opposition to railway Bills on even the minutest grounds and put the onus on proving compliance with standing orders upon the promoters rather than upon the party bringing a complaint of non-compliance. It was obligatory for the committees on standing orders to investigate and report on every allegation, and they did so with a conscientiousness that informed contemporaries found 'astonishing'.[46] The manner of conducting railway legislation through Parliament was the prime cause of the high cost of preliminary legal, administrative and parliamentary expenditure: the evidence is clear that throughout the 1830s the burden on railway promoters increased as new and more demanding requirements were instituted.[47] Opposition to Bills, of course, came from competing companies and also from landowners opposed to schemes.

A by-product of the system was that rather than face the prospect of opposed contests, which cost roughly four times unopposed ones,[48] and of future disputes over the value of land and the cost of arbitration and compensation, railway promoters bought off the opposition by paying over the odds for land. The extravagant and burdensome nature of the whole system of standing orders was challenged repeatedly in the 1840s and 1850s, and methods by which it could be modified to diminish the costs of railway legislation were discussed endlessly. No progress was made, however, because Parliament saw its prime responsibility as the protection of private property, through the system of standing orders, and was not prepared to move towards the system of 'expropriation'[49] practised abroad.

Even where railways did not succumb to the temptation of buying-off their opposition, Kellett has shown that the procedures laid down in the Land Clauses Consolidation Act and the Railway Clauses Consolidation Act of 1845 established the principle that the sums to be paid for land compulsorily acquired should be based upon its value to the vendor, together with further amounts as compensation for interruption of business, loss of goodwill, damages for the intersection and severance of property and for deterioration. To this sum was added a further 10 per cent for compulsory purchase, raised to 50 per cent in rural areas. The procedures, he notes, were an open invitation to landowners on the outskirts of towns to produce plans showing open fields or market gardens as potentially valuable building land. 'More exasperating still for railways were the verdicts sympathetic local juries returned on land without development prospects where compensation on the principle of loss of possible future gain was awarded.'[50]

The safeguarding of property thus went far, and from the records of independent valuations Kellett judges the railways paid twice the current market value. Much of the excess cost of land in Britain as compared with the Continent, then, arose from the institutional environment in which railway legislation was handled. Whereas in Europe the policy of governments was to regard the economic promotion of railways as a special duty, in Britain the framework was rooted in a contrasting attitude which prioritised the protection of private property and in the process enhanced its value by a third or more.[51] So although it is true to say that for some companies the high level of expenditure on land derived from quantity rather than price, it is also clear that in many cases the price was much higher than budgeted for. The 1839 select committee showed this to be true of railways in both England and Scotland, and of both rural and urban areas.[52]

The Board of Trade estimated in 1844 that on a comparative basis the excess capital cost of British railways due to parliamentary, legal and land costs was £2,700 per mile. This figure was framed as a cautious estimate; the important point is that in the comparative institutional sense the cost of preliminary expenses and of land and compensation was seen in the nineteenth century as a major cause of waste in the capitalisation of the system. As Laing noted in 1844, it was 'impossible to avoid contrasting the different systems adopted by the State with regard to railways in England and on the Continent. There, every encouragement has been given . . . in this country not only has no aid been given . . . but the restrictions imposed by the legislature have tended to increase very materially the cost of construction and to throw difficulties in the way of railway enterprise.'[53] Or, as a witness to the Royal Commission of 1865 put it, in Britain 'the various expenses of one class and another accompanying the passing of railway bills before a sod was cut would amount to something very nearly like the cost of the whole Prussian railway system'.[54]

The comparative exercise conducted by Laing extended also to analysis of the relative cost of constructing the works and stations and equipping the network with rails and stock. As regards the carrying establishment there was no great difference between English and contemporary European railways. The average of French and Belgian lines was £2,375 per mile; that of a sample of home railways, £2,500. The average cost of laying the permanent way, including rails, blocks and sleepers, at just over £4,150 per mile, was around £500 higher in England than in Europe, but stemmed from a tendency to use a heavier weight of rail.

Major relative excess costs in Britain derived firstly from the expense of stations, which averaged around £3,000 per mile compared with £1,000 per mile on the Continent, and secondly from the cost of engineering the system. In this respect British costs were not uniformly higher than Continental examples, and where there was a divergence in favour of the latter it was in some cases attributable to the nature of the terrain.⁵⁵ Nevertheless, there was a clear tendency for a large number of home railways to be engineered at higher cost than Continental lines constructed in broadly similar circumstances, and overall Laing estimated that excess relative costs in earthworks, bridges and other engineering works constituted about £5,000 per mile of railway. The causes were various. Having regard to the fact that some Continental lines were built by British contractors with British labour, and that the cost of labour employed in building railways was estimated by its efficiency as cheap in England as in France or Belgium, the view was that the excess of constructional costs reflected a pressure on companies to lay down lines with a view to avoiding parliamentary contests rather than getting the best route from an engineering viewpoint; also the impact on costs of standing orders which prohibited crossing roads on the level and imposed strict deviation restrictions on engineers.⁵⁶ A tendency to pursue engineering perfection rather than economy was apparent also, as was a tendency to rush into 'profuse expenditure' instead of limiting work to satisfy merely the existing demand for transport.⁵⁷ Overall, Laing's conclusion, drawn conservatively, was that the home railway system in 1844 could have been constructed at an average cost of £26,000 per mile rather than the £34,000 actually recorded in that year. The saving in national capital would have been equivalent to over 20 per cent of the paid-up capital of the railways, a sum sufficient to raise the gross return to over 9 per cent. This estimation he described as 'decidedly within the mark'. The figures in table 7, drawn directly from his work, confirm that Laing would have been justified in drawing a more severe conclusion based on a saving of £10,200 per route mile. The early disparity in capital costs between home and foreign railways was still apparent in 1865.⁵⁸

Table 7 *Laing's estimates of additional, capital expenses entailed in the construction of Britain's railways as compared with foreign countries in 1844*

Item	Excess cost per route mile of railway
Parliamentary expenses	£700
Legal and incidental expenses	£500
Land and compensation	£1,500
Permanent Way	£500
Stations	£2,000
Earthworks, bridges and engineering	£5,000

Source. Appendix to *Report from the Select Committee on Railways*, P.P., 1844, XI, appendix 2, No. 2, pp. 34–5.

VII

Since it seems beyond doubt that the over-capitalisation of home railways was established long before 1850, what do we make of the growth of capital per mile after 1870, even though it was much more modest than has generally been assumed?

The first point which bears repetition is that the increased capital was spent not on new lines but on extending existing capacity. The need arose partly from the pressure to improve the quality of the permanent way and general accommodation in response to public demand for higher standards. This is well established and accepted as a perpetual influence on railway investment in an era when the railways' obligations as a public service were becoming more clearly defined.[59] Equally, however, it arose from the fact that, far from becoming burdened with too much capital, many of the principal railways in the late nineteenth century were struggling to keep up with the demands of an expanding economy. To establish this we need to look back at the statistics of capital raised per route mile. As we saw, from 1850 to 1870 the amount of capital raised per mile was broadly constant. A new trend then set in, and by 1900 capital per route or track mile was enlarged as provision of extra track, sidings, standage and reconstruction of stations and warehouses took place. The statistics indicate that the break in trend came in the 1870s, and there is little doubt that the ferocity of the boom early in the decade so expanded the economy and the demand for transport that virtually all home railways were unable to keep up. The point was made quite explicitly by Lowthian Bell to the select committee on coal supplies in 1873.[60] There is a strong case for the view that that part of the excess capacity which derived before 1870 from creating capacity in advance of commercial needs[61] was more than swallowed up in the boom of 1870–73. From that time, despite odd and prominent examples of profitless investment, home railways, in periods of economic upswing especially, experienced some difficulty in coping with the demand for transport at key points on their systems, particularly as operating methods conditioned by public demand for better service led to lower train loads and more trains between 1870 and 1900. Reflecting this, the gross return on railway assets rose steadily, especially when the 'water' in capital accounts which has so confused some commentators is removed and the true picture obtained.

Support for this view comes from Sir Sam Fay of the Great Central, who in 1911 noted a constant growth of capital expenditure not on new lines such as his company's disastrous London extension but on the main trade routes, which were proving inadequate to carry the traffic seeking to find its way over them. The Great Central, he noted, was constantly having to spend to clear the main lines in the face of a constant growth of traffic in its traditional territories in industrial England.[62] This phenomenon was apparent also with London-based companies like the London & North Western and the Great Northern, which lacked the capacity to handle satisfactorily their expanding commuter traffic around the turn of the century.[63]

Thus we find, against some expectations, the gross earning power of railway assets increasing to a peak after 1900. Though the system remained heavily capitalised, and because of political constraints too much capital both fixed and working remained tied up in operating assets not used as productively as they might have been in different circumstances, the statistics indicate clearly that the burden of the early years was lessening. In these terms Ashworth's, Aldcroft's and Pollins's view that much of the high capitalisation of home railways arose after 1850 is quite simply wrong.

VIII

Our conclusions, then, are as follows.

Firstly, the course of railway capitalisation was a little different to that described by Hawke and Reed. Railway capital was subject to significant manipulation well before 1890, and existing statistics of capital formation must be revised in recognition of the fact. As regards the structure of the account, the effect of adjustment is to elevate the contribution of loans and debentures to financing both in England and Wales and in Scotland between 1870 and 1890. Adjustment highlights also the financial weakness of Scottish railways, where manipulation of shares commenced early and real capital formation was rather more sluggish from the early 1870s through to the early 1900s than has been suggested hitherto.

Secondly, review of the literary and statistical evidence demonstrates clearly that an initial high capital cost to Britain's railways was established by 1844, as Laing and others recognised. The excess relative to foreign railways derived from two sources: first, from the peculiarly high cost of promoting home railways and acquiring land; and, second, from the relatively high cost of earthworks, bridges and engineering, and from an insistence, quite clearly, on building 'ornamental station structures of a costly character where plain but substantial buildings would have answered',[64] sometimes also ahead of the initial demands of traffic. The excess capital of roughly £8,000 per mile identified by Laing in 1844 was due mainly to the engineering side, but over 35 per cent of that sum was nevertheless the direct result of the burdensome preliminary expenses and land charges incurred by home railways in establishing themselves. These figures do not fully support the railway view that *most* of the excess capital in the system was 'attributable to the outlay in obtaining statutory powers and in the acquisition of land' but they do show that the post-war tendency to dismiss this item as unimportant is incorrect. Capital accounts were seriously burdened in this respect; to dismiss it and the land question is to understate the institutional barriers to railway promotion and to misunderstand the entire sociopolitical background to railway development.

Finally, the tendency to deprecate the judgement of late nineteenth-century railway managers for their investment policies is shown to be misplaced. The adjustments to Hawke and Reed's estimates of capital growth modify the earnings ratios

The capitalisation of Britain's railways

for the 1870s and 1880s to a point where the gross returns on investment show steady growth from 1870 to 1914. Capital growth took place against a background of increasing demand for transport, and if net returns fell they can hardly be attributed to unremunerative investment in general, since the earning power of assets, once the 'water' is removed, was advancing constantly. Though there were individual failures, the argument that late Victorian railway policy led to general over-capitalisation is rejected decisively and the emphasis confirmed as lying on the operating front and on the factors affecting the conduct of traffic.

Appendix 1. Statement showing the amounts by which the capital of English and Welsh railways was nominally increased before 1890 on the conversion, consolidation and division of their stocks (£)

Year	Ordinary	Preference and guaranteed	Debentures	Cumulative total
1844	78,000[a]	–	–	78,000
1846	946,760[b]	–	–	1,024,760
1857	–	776,282[b]	–	1,801,042
1870	399,201[c]	–	–	2,200,243
1874	–	1,231,683[d]	125,822[d]	3,557,748
1875	–	156,420[e]	–	3,714,168
1876	–	20,800[e]	–	3,734,968
1877	–	296,408[f]	–	4,031,376
1878	1,149,643[b]	7,395,822[b]	160,989[g]	12,737,830
1879	–	1,703,568[b]	–	
	–	1,487,782[i]	–	15,929,180
1880	–	50,794[f]	–	15,979,974
1881	–	400,408[j]	–	
	–	1,096,574[g]	–	17,476,956
1882	–	192,205[k]	–	
	–	31,083[f]	–	
	–	478,375[l]	–	18,178,619
1883	–	100,000[f]	–	18,278,619
1885	–	129,757[m]	153,066	
	–	3,033,385[d]	–	
	63,266[n]	80,493[n]	108,120[n]	21,846,706
1887	2,640,914[j]	125,000[j]	–	
	840,599[o]	–	532,101[o]	25,985,320
1888	–	1,626,612[e]	–	
	–	824,560[i]	–	28,436,492

Year	Ordinary	Preference and guaranteed	Debentures	Cumulative total
1889	–	–	5,862,836[d]	
	2,764,284[a]	331,941[a]	207,147[a]	
	–	41,251[i]	511,896[p]	
	7,938[q]	244,980[q]	–	38,408,765
Total	8,890,605	21,856,183	7,661,977	38,408,765

Key and sources
a Taff Vale Railway: P.R.O. Rail 684/39 and 1110/453.
b LNWR: P.R.O. Rail 1110/269 and 270.
c NER: P.R.O. Rail 1110.
d Midland: P.R.O. Rail 491/8.
e GNR: P.R.O. Rail 236/51.
f MSLR: P.R.O. Rail 1110/158.
g LSWR: P.R.O. Rail 1110/283.
h LYR: P.R.O. Rail 1110/237.
i GER: P.R.O. Rail 1110/159 & 160.
j Metropolitan: P.R.O. Rail 1110/317.
k Alexandra Docks (Newport & South Wales) & Railway: P.R.O. Rail 1110/7.
l Furness: P.R.O. Rail 1110/147.
m Cambrian: P.R.O. Rail 1110/53.
n Cornwall Minerals Railway: P.R.O. Rail 1110/84.
o East London Railway: P.R.O. Rail 178/2.
p Devon & Somerset Railway: P.R.O. Rail 1110/98.
q Rhymney Railway: P.R.O. 1110/391.

Appendix 2. Statement showing the amounts by which the capital of Scottish railways was nominally increased before 1890 on the conversion, consolidation and division of their stocks (£)

Year	Ordinary	Preference and guaranteed	Debentures	Cumulative total
1870	–	103,500[a]	–	103,500
1873	1,254,815[a]	–	–	1,358,315
1874	–	42,820[a]	–	1,401,135
1876	2,783,659[b]	2,492,248[b]		
	2,422,485[a]	–	–	9,099,527
1878	–	189,864[a]	–	9,289,391

Year	Ordinary	Preference and guaranteed	Debentures	Cumulative total
1879	–	110,095^a		
		336,328^c	–	9,735,814
1881	–	987,770^d	–	10,723,584
1882	–	300,000^e	–	
		800,531^b	–	11,824,115
1883	424,280^c	229,382^c	50,301^c	12,528,078
1885	–	545,432^a		
		1,395^b		13,074,905
1887	–	175,941^d	–	13,250,846
1888	5,181,867^a	–	–	18,432,713
1889	–	713,760^d	–	19,146,473
Total	12,067,106	7,029,066	50,301	19,146,473

Key and sources.
a NBR: P.R.O. Rail 1110/354.
b Caledonian: P.R.O. Rail 1110/48.
c GNSR: P.R.O. Rail 1110/166.
d GSWR: P.R.O. Rail 1110/149.
e City of Glasgow Union: P.R.O. Rail 1124/128.

Appendix 3. Revised estimates of real capital raised by English and Welsh railways, 1870–89 (nominal additions excluded) (£'000)

Year	Share capital	Loans and debentures	Total
1870	319,290	118,642	437,932
1871	333,458	125,711	459,169
1872	343,863	127,576	471,439
1873	357,651	130,167	487,818
1874	371,166	134,003	505,169
1875	386,252	137,125	523,377
1876	400,570	140,558	541,128
1877	409,102	144,744	553,846
1878	420,372	146,278	566,650
1879	426,552	150,675	577,227
1880	434,066	152,158	586,224
1881	443,526	154,004	597,530
1882	457,256	158,323	615,579
1883	465,437	164,265	629,702

Year	Share capital	Loans and debentures	Total
1884	475,441	168,369	643,810
1885	479,748	172,618	652,366
1886	488,060	174,758	662,818
1887	493,562	177,207	670,769
1888	494,921	183,303	678,224
1889	496,878	179,239	676,117

Source. Hawke and Reed, op. cit., modified by appendix 1.

Appendix 4. Revised estimates of real capital raised by Scottish railways, 1870–89 (nominal additions excluded) (£'000)

Year	Share capital	Loans and debentures	Total
1870	45,525	16,874	62,399
1871	46,847	17,321	64,168
1872	48,938	17,659	66,597
1873	49,400	18,367	67,757
1874	51,050	18,865	69,915
1875	52,782	18,693	71,475
1876	54,707	18,900	73,607
1877	56,609	19,205	75,814
1878	58,192	19,698	77,890
1879	61,181	19,767	80,948
1880	62,243	20,343	82,586
1881	63,509	20,576	84,085
1882	65,263	20,652	85,915
1883	65,239	20,752	85,991
1884	69,531	21,504	91,035
1885	68,242	21,703	89,945
1886	68,888	22,622	91,510
1887	70,520	23,178	93,698
1888	72,318	23,358	95,676
1889	73,445	23,381	96,826

Source. Hawke and Reed, op. cit., modified by appendix 2.

Notes

1 G. R. Hawke and M. C. Reed, 'Railway capital in the United Kingdom in the nineteenth century', *Economic History Review*, second series, XXII (1969).
2 *Ibid.*
3 W. Vamplew, 'Nihilistic impressions of British railway history', in D. McCloskey (ed.), *Essays on a Mature Economy: Britain after 1840* (1971), p. 355. Also R. J. Irving, *The North Eastern Railway Company, 1870–1914: an Economic History* (Leicester, 1976), pp. 140–1.
4 See *Select Committee on Caledonian Railway (Conversion of Stock) Bill*, 1890, qq. 525–9 (Giffen), held as P.R.O. Rail 1124/128.
5 *Ibid.*
6 P.R.O. Rail 684/39.
7 P.R.O. Rail 1110/269; also P.R.O. Rail 1063/148 for the Act of Parliament sanctioning the arrangement of capital.
8 The consolidation of North Eastern Railway ordinary stock in 1870 is a further major example of this process. See Irving, *op. cit.*, pp. 17–18.
9 *Statement showing the amounts by which the capital of the railway companies of the United Kingdom have been nominally increased on the conversion, consolidation and division of their stocks*, Parliamentary Papers, 1890/91, LXXV.
10 See *Select Committee on Caledonian Railway (Conversion of Stock) Bill*, above, qq. 525–9.
11 *Ibid.*
12 *Ibid.*
13 See variously E. D. Chattaway, *A Rudimentary Treatise on Railways: their Capital and Dividends* (1855–56), p. 22; W. R. Lawson, sometime chairman of the English Railway Shareholders' Association, giving evidence to the *Departmental Committee of the Board of Trade on Railway Agreements and Amalgamations*, P.P., 1911, XXIX, qq. 1183–90. Also the summary of views in the Labour Research Department's *Labour and Capital on the Railways* (1923), chapter 1.
14 Chattaway, *op. cit.*, for example.
15 Major examples are the *Select Committee on Railway Communication*, P.P., 1839, X; *Select Committee on Railways*, P.P., 1844, XI, especially Appendix to the Report by S. Laing, Chief Law and Corresponding Clerk to the Board of Trade; *Select Committee on Railway Acts Enactments*, P.P., 1846, XIV, and *Select Committee on Railway and Canal Bills*, P.P., 1852–53, XXXVIII.
16 *Report of the Board of Trade Railway Conference*, P.P., 1909, LXVII, item 10, *The Acquisition and Holding of Land*. Also J. Francis, *A History of the English Railway* (1851); H. Spencer, *Railway Morals and Railway Policy* (1855); and Lawson, *op. cit.*
17 For example, C. E. R. Sherrington, *The Economics of Rail Transport* (1928); J. H. Clapham, *An Economic History of Modern Britain: the Early Railway Age* (1938), p. 388; W. V. Wood and J. Stamp, *Railways* (1928), p. 12; L. C. A. Knowles, *The Industrial and Commercial Revolutions in Great Britain during the Nineteenth Century* (1930), p. 257.
18 H. Pollins, 'A note on railway constructional costs, 1825–50', *Economica*, XIX (1952).
19 W. Ashworth, *An Economic History of England, 1870–1939* (1960), chapter 5.
20 D. H. Aldcroft, *British Railways in Transition* (1968), p. 12, and (with H. J. Dyos) *British Transport* (1969), p. 195.
21 D. H. Aldcroft, 'The efficiency and enterprise of British railways, 1870–1914', *Explorations in Entrepreneurial History*, V (1968).
22 H. Pollins, *Britain's Railways: an Industrial History* (1971), p. 114.
23 M. Robbins, *The Railway Age* (1962), chapter 4.
24 C. Savage, *An Economic History of Transport* (1959), chapter 2.
25 P. S. Bagwell, *The Transport Revolution since 1770* (1974), pp. 100–2.
26 T. R. Gourvish, *Mark Huish and the London & North Western Railway: a Study of Management* (Leicester, 1972), chapter 1; also his *Railways and the British Economy, 1830–1914* (1981), p. 16.
27 J. R. Kellett, *The Impact of Railways on Victorian Cities* (1969), p. 11 and appendix I, pp. 427–31.
28 R. J. Irving, 'The profitability and performance of British railways, 1870–1914', *Economic History Review*, second series, XXXI (1978).
29 *Ibid.*
30 B. R. Mitchell, 'The coming of the railway and United Kingdom economic growth', *Journal of Economic History*, XXIV (1964), table 2.
31 Clapham, *op. cit.*, p. 385.
32 Estimates derived from R. Fishlow, *American Railroads and the Transformation of the Antebellum Economy* (Harvard, 1965), pp. 316 and 399; *Historical Statistics of the United States* (Washington, D.C., 1960), p. 428, and B. Z. Ripley, *Transportation* (New York, 1902).
33 R. O. Christenson, 'The State and Indian railway performance', part 1, *Journal of Transport History*, third series, II, 1981.
34 P. J. Cain, 'Private enterprise or public utility? Output, pricing and investment on English and Welsh railways, 1870–1914', *Journal of Transport History*, third series, I, 1980.
35 See *Select Committee on Railway and Canal Bills*, P.P., 1852–53, XXXVIII, q. 79 (Laing) and q. 1043 (Stephenson). See also Gourvish (1973), *op. cit.*, p. 23.
36 *Royal Commission on Railways*, P.P., 1867, XXXVIII, q. 16183, (Eborall).

37 R. M. Stephenson, *A Rudimentary Treatise on Railways and Sketches of the Construction and Material* (1850), p. 8.
38 P.P., 1909, LXVII, item 10, *op. cit*. Also Lawson, *op. cit*.
39 J. Simmons, *The Railway in England and Wales, 1830–1914* (Leicester, 1978), pp. 43–4.
40 Kellett, *op. cit.*, and Simmons, *op. cit.*
41 See *Select Committee* of 1839, *op. cit*., q. 858 (Saunders), q. 3749 for the Newcastle & Carlisle and q. 1443 for the London & Croydon Railway.
42 See *Appendix to the Report from the Select Committee on Railways*, P.P., 1844, *op. cit.*
43 The relationship between constructional costs and rates and fares was a common topic for discussion before inquiries in the 1840s.
44 *Select Committee* of 1844, *op. cit.*, appendix 2, p. 32.
45 *Ibid.*
46 See especially *Report from the Select Committee of the House of Lords on the Management of Railways*, P.P., 1846, XIII, and evidence therein, notably qq. 179–83 and 625.
47 *Ibid.*, q. 758.
48 See *Report of the Select Committee Appointed to Enquire into the Best Mode of Securing the Public Interest and Diminishing Parliamentary Expenses in Reference to Railway and Canal Legislation*, P.P., 1857–58, XIV, q. 151.
49 *Ibid.*, q. 1167.
50 J. R. Kellett, 'Urban and transport history from legal records', *Journal of Transport History*, first series, VI, 1963–64. Also his *The Impact of Railways on Victorian Cities*, *op. cit.*
51 *Select Committee on Railway Communication*, P.P., 1839, *op. cit.*, qq. 3444; also Kellett, above, *Journal of Transport History*, 1963–64, and *The Impact of Railways on Victorian Cities*, p. 392.
52 Witnesses from the GWR, the London & Southampton, the LBSCR, the Arbroath & Forfar and Newcastle & Carlisle railways all testified to this effect, which in some cases operated alongside the need to buy more land than allowed for. See also Kellett, *op. cit.*, chapter 12.
53 *Select Committee* of 1844, *op. cit.*, appendix 2, p. 36.
54 *Royal Commission on Railways*, P.P., 1866, *op. cit.*, q. 1155.
55 *Select Committee of the House of Lords*, P.P., 1846, *op. cit.*, qq. 483–513 (Stephenson).
56 See evidence of J. Baxendale and W. A. Wilkinson to the *Select Committee on Railways*, P.P., 1844, *op. cit.*, qq. 3235–9 and 3872.
57 See Chattaway, *op. cit.*, p. 130.
58 *Royal Commission on Railways*, P.P., 1866, *op. cit.* appendices D, K and M.
59 See for an early identification of the situation *ibid.*, qq. 15289, evidence of T. E. Harrison. Also Irving, *op. cit.*
60 See *Select Committee on the Present Dearness and Scarcity of Coal*, P.P., 1873, X, q. 6170 (Bell).
61 See Chattaway, *op. cit.*, also W. T. Jackman, *The Development of Transportation in Modern England* (1962 edition), p. 601.
62 *Departmental Committee* of 1911, *op. cit.*
63 See R. J. Irving, 'British railway investment and innovation, 1900–14', *Business History*, XIII (1971).
64 Chattaway, *op. cit.*

Acknowledgements

A first draft of this paper was read to participants in the Transport History Group Conference at the City of Birmingham Polytechnic in September 1981. The author acknowledges the value of helpful criticisms offered. Thanks are due to the Public Record Office and the University of York for access to materials held in their possession, and to the editors of this journal for their assistance in the preparation of the paper for publication.

7 The state and Indian railway performance, 1870–1920:
Part 1: Financial efficiency and standards of service

R. O. CHRISTENSEN

The first miles of railway track in India were opened in 1853. By 1920 the network had been extended to a total of over 36,000 route miles, and, at least by this criterion, India had become one of the half-dozen greatest railway powers in the world.

Most writing on the Indian railway system has been concerned chiefly with its consequences for the economic and social well-being of the sub-continent, with debate following the well known division between those who saw British enterprise in India as a stimulus to economic growth and those, notably nationalists, who took a less optimistic view. Official opinion, as expressed in a memorandum published in 1909, was that the railways had brought about an 'incalculable benefit' to the Indian economy and society, with the net gain to traders, producers and the travelling public estimated at about £100 million per annum.[1] Critics, however, argued that the railways were an extravagance unsuited to Indian needs, and that, having been built mainly to serve the commercial, administrative and strategic purposes of the British, they failed to generate economic development within the sub-continent itself.[2] Nevertheless, despite the attention that has been given to the question of the wider developmental implications of the railway system, a thorough analysis of its consequences for the Indian economy, along the lines undertaken by Fogel and Hawke for the United States and Britain, has yet to be made.[3]

Very little, however, has been written on the performance of the system itself, and in particular on the reasons for the growing dissatisfaction with its functioning in the decades before 1920.[4] For whatever effects, beneficial or otherwise, the railways may have had on Indian economic and social life, by the beginning of the twentieth century there was general agreement that the system itself was far from perfect. Under pressure of growing discontent, the government commissioned a series of reports on its operation and administration. The last and most critical, that of the Acworth Committee, published in 1921, found little difficulty in concurring with Indian public opinion as to 'the entire inadequacy of the Indian railway system to meet the needs of the country'.[5] Among the principal deficiencies with which it dealt were widespread congestion, embargoes on goods traffic which were seriously hindering commercial activity, and the overcrowding suffered by third-class passengers. The report as a whole was a strong condemnation of past railway policy, and stressed unequivocally the necessity for urgent measures of reform, especially regarding capital funding and administrative structure.

Although the problems examined by the Acworth Committee had become particularly evident during and immediately after the first world war, their origins may be found as far back as the abandonment of the 5 per cent guaranteed-interest terms in 1870. The original guarantee system had, certainly, given rise to difficulties of its own, above all that deficits on the interest account constituted a growing burden on State revenues. But as the State came increasingly to take over and develop the railways on its own account, problems of a different kind began to emerge. From 1870 there evolved what appeared to be a highly complex system of management, comprising, by the early twentieth century, a large number of separate administrative bodies of several distinct types: lines owned and worked directly by the State; lines owned by the State and worked by private companies, usually guaranteed; lines owned and worked by guaranteed companies; lines owned by companies which were rebate-aided or otherwise subsidised; and (a rather small proportion) lines owned by independent, unassisted companies.[6] The essential point, however, is that gradually the State had acquired ultimate financial responsibility for almost the entire railway network, and had ownership of over 90 per cent of the total route mileage, including all the major trunk lines.

The concern of this paper is to examine how far, and in what respects, State ownership and control affected the performance of the railways, or allowed them to operate fully as a commercial enterprise. The first part will deal with some of the criteria applied in measuring the efficiency of the Indian railway system, and the extent to which they were compatible with each other. The second part, to be published in the next issue of this journal, will consider the role of the government with regard to rating policy and the financing of capital expenditure, and the relative importance of these factors in determining railway development and performance.

I

What yardsticks should be employed to judge the efficiency of a railway system? It should be said that the concept of efficiency is to a considerable extent problematical, since there are several indicators, including financial results, traffic handling, productivity and standards of service, which may be taken to define it. Not all are equally susceptible of measurement; nor do they necessarily point in the same direction. Efficiency is not altogether an indivisible concept, and an improvement or decline as shown by one indicator does not always show itself in a corresponding movement by others.

The question of efficiency appears to present a particular set of difficulties in relation to the Indian railway system during the half-century after 1870, in so far as this was a period of continual extension of the network, with an average of over 600 route miles being added each year. Most of the additional mileage comprised subsidiary and branch lines, and many of these were constructed not for commercial reasons but for military/strategic purposes or to afford protection to famine-prone districts. Nevertheless, although extensive construction and the non-commercial basis of many lines might imply that criteria generally used to evaluate performance need to be modified, the structure of the railway network suggests that these factors were not as significant as they seem. The most important components of the system were the trunk lines linking the main ports of shipment with inland commercial centres and agricultural districts. These arterial routes, most of which had already come into operation by 1870, carried the bulk of the traffic (the nine principal commercial lines carried over 80 per cent of total traffic in the 1880s, and nearly 75 per cent thirty years later), and it was their performance which very largely determined that of the system as a whole.

Perhaps the simplest way to approach the question of the relative efficiency of the Indian railways is to compare their performance with that of railways in other countries during the same period. A comparative method was, indeed, adopted by certain contemporary writers on the subject. Thomas Robertson, for example, in his 1903 report criticised several aspects of Indian railway performance by reference to American standards, arguing that 'the conditions of the traffic in America and India are very similar'.[7] Certain qualifications are necessary, however. While superficially freight traffic patterns in India resembled those in the USA, consisting mainly of bulk produce carried over long distances, the size of consignments in the two countries was very different, those in America being much larger.[8] Other factors have also to be taken into account. Thus in India wage rates for the largest grades of labour were considerably below those in industrial countries, between one-fifth and one-sixth those prevailing in Britain. The institutional framework and its influence on tariffs or railway investment decisions was in important respects different from that elsewhere. So, too, was the general economic environment in which the Indian railways operated; while a monsoon failure could seriously affect commercial activity, exogenous factors of this kind were of relatively little significance in Europe or America. Any detailed comparison of the performance of the Indian railways with that of other systems has, therefore, to be undertaken with care.

The most general, and most commonly used, indicator of relative efficiency is the operating ratio, or the ratio of working expenditure to gross receipts. In many respects the operating ratio is an unsatisfactory measure of railway performance, since it is liable to be affected by both supply and demand factors, such as changes in labour or fuel costs, which may be entirely unrelated to actual working efficiency.[9] Although improvements in operating methods may be reflected in the relationship of expenses to earnings, the operating ratio should not be seen as anything other than a broad guide to financial efficiency.

With regard to the British railways, it is well known that they showed a steady deterioration in operating ratio from around 50 per cent in 1860–80 to over 60 per cent in the years before 1914, due partly to the fact that the companies were caught between rising factor costs and fixed freight charges, but also to a form of monopolistic competition between them which led to greater overhead expenses for the provision of improved services.[10] In the USA the situation was even more serious, as the ratio rose from around 65 per cent in the 1890s to over 70 per cent before 1914, largely because of intense competition between the American companies, which cut considerably into profit margins and led many into bankruptcy.

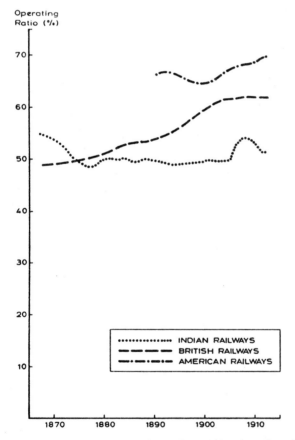

Fig. 1 The operating ratios of the Indian, British and American railway systems (five-year moving averages). Sources. *Railways in India*, annual administration reports; *Railway Returns of the Board of Trade* (annual); *Historical Statistics of the United States* (Washington, D.C., 1960), Series Q 114.

In India, by contrast, the operating ratio of the railway system as a whole showed a marked degree of constancy, and was low by contemporary standards. Once initial outlay on the main trunk lines had been completed in the early 1870s, it was usually around or below 50 per cent.[11] The only exception before 1914 was a rise in the ratio to an unprecedented peak in 1908 of over 60 per cent from less than 48 per cent in 1905, but within four years it had reverted to its usual level. The 1908 result did not, however, signify any deterioration in operating performance, being almost entirely attributable to a heavy falling-off in goods receipts, which in turn was due mainly to a decrease in exports of wheat to Europe in consequence of scarcity and high prices in India.[12] Only after the first world war was there a sharp upward movement of the operating ratio, to a peak in 1921–22 of 76·2 per cent, but this was wholly accountable to the after-effects of the war and the economies in expenditure which it had imposed.

The generally very favourable comparison of the Indian operating ratio with that of the British and American railways may be explained in various ways. Few of the elements of competition seen in the USA and, to a lesser extent, in Britain were present in India. Many of the Indian lines enjoyed a complete regional monopoly, and even where this was partially eroded competition never approached the intensity of the rate wars seen in America. In addition, very low wage levels for semi- or unskilled labour gave the Indian system a considerable advantage. Yet this may simply point to one of the weaknesses of the operating ratio as an indicator of efficiency, mentioned above. Especially in a country in which labour is a cheap factor, a low level of expenditure in relation to receipts does not necessarily say much about working efficiency. The question, therefore, is how far a good operating ratio adequately reflects Indian railway performance in such aspects as traffic operation and standards of service.

II

As the immense body of official literature and statistical material shows, there was no lack of interest on the part of management or government in the efficiency of the system during the period 1870–1920. While reflecting the government's direct financial responsibility for the railways, this high level of concern also led to some significant improvements in traffic operation. Good financial results and good operating performance were, indeed, regarded as closely connected.

Probably the most important improvement was in goods train loading, as a consequence of new methods of statistical analysis introduced in the early 1870s, mainly through the efforts of A. M. Rendel, Consulting Engineer to the East Indian Railway. The Indian railways were among the first in the world to employ ton- and passenger-mile statistics, and on the basis of the practical results which they afforded Juland Danvers, then Government Director of the Indian Railway Companies, strongly and repeatedly argued for their superiority over the train-mile statistics at that time (and, indeed, for the following four decades) in use in Britain as the chief measure of traffic operations. The proportion of train miles run to the length of line in operation, he maintained, was no real test of the efficiency of working; nor were train-mile receipts an accurate guide to profitability. Ton-mile results alone could provide a proper measure of actual work done.[13]

Rendel's findings demonstrate the way in which these statistics could be applied. In 1868 it was noted that the train-mile expenses on several lines, notably those operating from Bombay, were comparatively high. This was attributed mainly to fuel costs. Thus the EIR, which ran from Howrah to Delhi and to which Bengal coal was readily available, was paying only 20s per ton for coal, while the Bombay lines, which were dependent on English or Australian coal, were paying between 60s and 80s per ton.[14] It was confidently expected that train-mile operating costs on these lines would fall once it became possible to transport the cheaper Indian coal to them from Bengal. Rendel reached a somewhat different conclusion, however. The question of coal, he argued, was of marginal importance. Ton-mile costs were relatively low on the Madras railway, which was at a greater disadvantage than any other line with respect to coal supplies. Moreover the cost of carriage was not necessarily lowest on lines with relatively large traffic; the Madras railway had the least traffic of all the major lines. Rendel found, instead, that the principal reason

Table 1 *Ton-mile receipts and expenses of the main Indian lines, 1872 (pence)*

	East Indian Railway (Howrah –Delhi)	Great Indian Peninsula (Bombay)	Bombay Baroda & Central India	Madras
Average charge per ton mile	1·24	1·59	1·50	1·45
Average cost per ton mile	0·41	0·97	1·00	0·67
Average profit per ton mile	0·83	0·62	0·50	0·78

Source. *Railways in India*, Parl. Papers, 1872, XLIV, appendix prepared by A. M. Rendel.

for generally high working costs on the Indian railways was the insufficient loading of trains, and the consequent unnecessary mileage run. On the line showing the best operating performance, the EIR, the average freight train load was only 109 tons, while on other lines it was even lower, despite the fact that much of the traffic comprised export consignments carried over long distances. On the basis of his ton-mile results Rendel showed that loads could be increased to an average of between 130 and 140 tons on all lines, while costs could be more than halved.[15]

Comprehensive ton-mile statistics for the Indian railway system as a whole were published from 1873 onwards, and marked improvements in freight operation followed. By the early 1880s the average goods train load on all Indian lines had increased to about 115 tons, and by the turn of the century exceeded 200 tons.[16] Better loading contributed to the lowering of freight rates during the period, especially for coal and food-grain traffic. As Danvers noted:

> When, by means of these results [ton-mile statistics], it was found that there was a waste in the train miles run and in the train loads, measures were taken to prevent the extravagance. The number of wagons in a train was increased, and the load in each wagon was made greater. The consequence has been that the average loads by train and by vehicle have been larger. Economy has been secured, and the public have benefited; for the railway authorities found that it was expedient and desirable to reduce rates.[17]

For Danvers the experience of the Indian railways provided a clear lesson for their British counterparts: more informative statistics were a necessary precondition for securing greater overall efficiency. Even though traffic patterns in Britain differed in important respects from those in India (freight hauls were very much shorter and consignments smaller), goods train loads in the two countries were still in striking contrast to each other. To a great extent this could be attributed to the fact that few British railway companies employed ton-mile statistics before the first world war.[18]

The improvements in goods loading achieved by the Indian railways are reflected in overall productivity trends. Although a detailed analysis of changes in labour productivity is made difficult by the absence of data relating to each railway department, the aggregate movement from the 1880s was clearly upwards. Between the quinquennia 1881–85 and 1906–10 the number of train miles run per man showed an average rate of increase of about 0·6 per cent per annum; if ton miles are taken as the measure of output an annual growth rate of over 1·2 per cent is indicated (table 2). These figures confirm the importance of improvements in loading in bringing about greater operating efficiency. Again, too, the Indian railways compare favourably with those in Britain, whose overall productivity was almost stationary during this period.[19]

Given these improvements in traffic operation, however, it seems slightly paradoxical that the Indian railway system should have encountered increasing difficulties from the late nineteenth century. Part of the answer is that such problems as congestion were not easily deducible from most of the statistical indicators available. The innovations introduced by Rendel and Danvers in the 1870s were not developed further or in such a way as to show, for example, efficiency over time through the measurement of ton miles carried per working day. A committee set up in 1880 to revise and standardise the collection of statistics, although briefly acknowledging the utility of ton-mile results, made no significant recommendations beyond a simplification and enlargement of existing returns.[20] Thereafter the published statistics underwent a steady expansion, but no further review of their collection or of methods of analysis was made until 1923. In volume the Indian railway statistics

Table 2 Productivity indices of the Indian railways, 1871–1914
(1881–85 = 100)

Period	Train miles per man	Ton miles per man
1871–74	100·54	—[a]
1876–80	96·64	84·75
1881–85	100	100
1886–90	100·17	100·47
1891–95	113·69	117·20
1896–1900	113·57	122·23
1901–05	116·42	132·61[b]
1906–10	118·81	139·46
1911–13/14	122·70	172·42

Notes: (a) insufficient data available; (b) excludes 1903 and 1904, for which years ton-mile results are not given. Number of employees in 1975 not stated.

Source. Based on returns in *Railways in India*, annual administration reports to Parliament.

were greater than those of any other country in the world, but as the Acworth Committee's report dryly remarked, 'Their value is not proportionate to their volume.'[21]

III

The first general examination of the problem of the unsatisfactory working of the Indian railways was made in the Robertson report of 1903. The approach adopted by its author, who had travelled extensively on the railways of the USA, was to compare the standards of service of the Indian and American systems; on this basis he found that the performance of the former was much below what it should have been 'in a country which has had the benefit of railway communications for nearly fifty years'.[22] As has already been noted, however, a comparative method of analysis is open to certain objections. One Indian railway official responded to Robertson's line of argument by roundly criticising him for his failure to take into account the special historical conditions under which railway development had taken place in India, the incomplete spread of the network over the subcontinent, and the nature of the traffic offering.[23] Nevertheless, while Robertson's approach may in some respects have been defective, he indicated clearly the difficulties of the Indian system which were to become increasingly evident in the following two decades.

The specific problems he examined included slow train speeds, poor provision for third-class passengers, and delays in goods traffic resulting from inadequacy of rolling stock. All were greatly exacerbated by the rapid growth of both goods and passenger traffic from the 1890s,[24] the consequence being a vicious circle in which overcrowding and congestion became progressively worse. Even the fastest mainline passenger train in India ran at little over half the average speed of its equivalent in the USA, and the protracted length of time of journeys served only to hold down the frequency of running and thus raise the average number of passengers per train, already higher than in any other country in the world.[25] It was widely acknowledged that the largest class of passengers had to endure intolerably overcrowded and insanitary conditions of travel. Some of the worst were brought to the attention of the Supreme Government in 1906:

> It is not an infrequent sight to find the passengers huddled together in trucks and waggons intended for carrying goods. These goods waggons have labels attached to them limiting the number of passengers to be accommodated in the hot and cold weather, but this injunction is honoured more in the breach than in the observance.[26]

The Acworth Committee received similar evidence; one witness instanced the case of a train providing on average accommodation for 297 passengers and carrying an average of 842 passengers per journey.[27] Freight traffic fared no better. Goods train speeds on main lines recorded by Robertson ranged from an average of 12 m.p.h. for the 954 mile journey from Delhi to Howrah, to 3·7 m.p.h. for the 466 miles from Howrah to Lucknow. On shorter hauls they were lower still, as low as 2 m.p.h. for the line between the Bengal coalfields and Calcutta.[28] Despite embargoes placed on several classes of traffic, delays became longer. Coal and cotton were among the major commodities whose export was held up, for as long as six months, while the Tata Iron & Steel Company estimated that in 1920 as much as 5 per cent of its production had been lost as a result of railway congestion.[29]

To a certain extent these difficulties could be explained by bad working and administrative methods. Train speeds, for example, were considerably slowed by the time allowed for stops, while goods loading suffered from a lack of proper planning. The fundamental problem, however, lay with the inadequacy of capital equipment, especially rolling stock, to meet the demands of growing traffic. Robertson emphasised in particular the failure to replace outdated wagons with the high-capacity bogie vehicles then widely in use in the USA. At the turn of the century goods train loads of 3,000 tons or more were common in America, but in India the best load was no more than 1,000 tons, and this was well above the average.[30] Thus although marked improvements in loading on existing stock were achieved, in relative terms the Indian railways were falling behind. Moreover, the use of four-wheeled trucks rather than open bogies meant that both dead weight and the length of trains were far greater with respect to paying load than necessary. Even if differences between Indian and American traffic patterns are taken into account,[31] there was still much scope for the introduction of more advanced rolling stock on the most important trunk lines. The congestion resulting from the general shortage of carrying capacity was made worse by other deficiencies of capital equipment, such as small and badly laid out marshalling yards and delays in the provision of automatic brakes and safety devices. In short, the failure either to renew or to extend the capital stock of the Indian railways in line with the traffic offering was the main factor responsible for the serious decline in standards of service which became evident from the beginning of the twentieth century.

The problem of inadequate renewal of capacity was not confined to the Indian railways. In Britain there was a similar failure to undertake improvements in capital equipment, affecting both the efficiency of traffic operation and, in consequence, profitability.[32] The reasons for this apparent neglect in the two countries were not identical, however. There is no evidence, for example, that the fragmented traffic patterns and division of ownership of rolling stock which delayed costly innovations in Britain affected the Indian lines. Also, while in Britain the main railway construction phase was over by the 1870s, in India new trunk lines continued to be built until early in the twentieth century, with several thousand miles of branch and feeder lines added before the interruption of war. Arguably the interrelated costs arising out of the introduction of improvements should have been proportionately lower on a system still in the course of extension than on one already completed. But the opportunity was never realised. Instead, extensions of the system appear to have been undertaken largely at the expense of improvements.

The contrast between the Indian and the British railways with regard to the problem of unimproved capital equipment is perhaps most clearly brought out by comparing their respective capital/labour ratios. If the British railways were becoming increasingly labour-intensive, both over time and in relation to other systems, such as that of the USA, which received considerable investment in more advanced equipment, by Indian standards they were nevertheless highly labour-intensive. The number of employees per unit of capital investment in India was consistently well over four times that in Britain (table 3). The figures for labour productivity show a similar differential. Although traffic movements per man increased in India during the period 1870–1914, while in Britain they were stagnant, they were still more than five times better on the British railways in the years immediately prior to the war.

The very labour-intensive pattern of working in India may probably be explained largely by the low level of wages prevalent there. Over 95 per cent of the total work force were Indians, most of them engaged in semi- or unskilled occupations, whose rates of pay were greatly below those of their counterparts in Europe or America. In the absence of trade union organisation, pressure to raise wages before 1914 was slight. Some rises occurred from the end of the nineteenth century, when mills and other industrial concerns began to compete for labour, but the effect on the railways' financial position was small. In these circumstances it may have been economically quite rational not to have replaced labour with relatively costly capital

Table 3 *The ratio of labour to capital, Indian and British railways*

Year	Employees per £ million of capital investment		Column 1 as a multiple of column 2
	1 India[a]	2 UK[b]	
1871	732	174	4·2
1881	1259	211	6·0
1891	1571	231	6·8
1901	1442	266	5·4
1911	1667	280	6·0

Notes
(a) Figures for capital outlay on the Indian railways are given in the source in Rs. crores (one crore = ten million); on conversion, see note to appendix.
(b) Capital investment in Britain has been taken as total capital paid up and raised by loans; figures for employees are rounded to thousands in the source. The results in both columns are necessarily only very approximate.
Sources. Based on *Railways in India*, annual administration reports; *Railway Returns of the Board of Trade*; and B. R. Mitchell and P. Deane, *Abstract of British Historical Statistics* (Cambridge, 1962), p. 60.

equipment. Perhaps even more important, however, was that the Indian railways were greatly under-capitalised because of a shortage of investment funds, a theme that will be taken up in Part II of this paper.

Finally, the operating deficiencies resulting from inadequate and outdated capital stock did not affect, and were not reflected in, financial performance. Unlike the British railways, those in India did not have to face a combination of sharply rising costs and falling or fixed charges. Expenses per train mile held steady, while earnings rose in proportion to traffic carried. What the qualitative evidence and comparative tests suggest, therefore, is that though general financial indicators, particularly the operating ratio, show that the Indian railway system functioned relatively well, these indicators were misleading in so far as they failed to take into account constraints to which the system was subject. A good operating ratio may, indeed, have been achieved precisely at the cost of efficiency in other respects.

Ultimately, then, such improvements as better goods loading and a rise in productivity could have only limited results. By the turn of the century, caught between a rapid growth of traffic and inadequate carrying capacity, it became increasingly difficult for the railways to maintain standards of service. The concept of railway service was not, however, well developed in India, even though, as the Acworth Committee report testifies, the deterioration before 1920 led to widespread public dissatisfaction. But while in Britain, under the influence of pressure groups, the railways came to be regarded largely as a public service, to be operated in the interests of traders and the travelling public,[33] in India railway policy appears instead to have been determined much more by financial considerations. Some of the reasons for this concern with financial efficiency will be examined in the second part of this paper, which will deal with aspects of government railway policy.

[To be concluded

Appendix

Main financial results of the Indian railway system, 1870–1920 (Rs. crores[a])

Year	1 Total capital outlay[b]	2 Gross receipts	3 Working expenditure	4 Net receipts	5 Operating ratio (%) (3/2)	6 Net receipts as % of capital outlay (4/1)
1870	90·00	6·67	3·63	3·04	54·47	3·37
1871	90·01	6·59	3·68	2·91	55·82	3·24
1872	90·01	6·83	3·73	3·10	54·68	3·44
1873	91·73	7·23	3·78	3·45	52·28	3·76
1874	95·87	8·34	4·04	4·30	48·44	4·48
1875	100·96	7·91	3·97	3·94	50·23	3·90
1876	104·78	9·34	4·46	4·87	47·81	4·65
1877	109·04	12·11	5·39	6·73	44·47	6·17
1978	118·30	11·25	5·62	5·63	49·97	4·76
1879	122·33	12·08	6·26	5·82	51·84	4·76
1880	128·57	12·87	6·48	6·39	50·37	4·97
1881	140·81	14·32	7·07	7·25	49·37	5·16
1882	143·24	15·35	7·67	7·68	49·95	5·36
1883	148·31	16·39	7·97	8·42	48·62	5·68
1884	155·78	16·07	8·16	7·91	50·76	5·09
1885	162·19	18·03	8·89	9·14	49·30	5·64
1886	170·50	18·70	8·93	9·77	47·75	5·73
1887	182·88	18·47	9·10	9·36	49·31	5·12
1888	193·04	19·76	9·87	9·89	49·96	5·12
1889	205·05	20·49	10·38	10·12	50·64	4·93
1890	213·67	20·67	10·30	10·36	49·87	4·85
1891	221·06	24·04	11·30	12·74	47·02	5·76

Year	1 Total capital outlay[b]	2 Gross receipts	3 Working expenditure	4 Net receipts	5 Operating ratio (%) (3/2)	6 Net receipts as % of capital outlay (4/1)
1892	226·76	23·20	10·89	12·31	46·94	5·42
1893	232·67	24·06	11·33	12·73	47·12	5·46
1894	237·24	25·48	11·97	13·51	46·97	5·70
1895	243·76	26·20	12·10	14·10	46·20	5·78
1896	268·27	25·32	12·12	13·20	47·86	4·92
1897	281·49	25·56	12·46	13·11	48·72	4·66
1898	292·17	27·41	13·00	14·41	47·37	4·94
1899	308·58	29·37	13·94	15·43	47·43	5·01
1900	329·61	31·54	15·10	16·45	47·85	4·99
1901	339·17	33·60	15·72	17·88	46·79	5·27
1902	349·77	33·93	16·70	17·22	49·24	4·92
1903	341·11	36·01	17·11	18·90	47·52	5·54
1904	347·91	39·67	18·79	20·88	47·36	6·00
1905	358·26	41·70	19·95	21·75	47·85	6·07
1906	371·01	44·14	22·02	22·11	49·89	5·96
1907	391·87	47·31	24·32	22·98	51·42	5·86
1908	411·92	44·83	27·00	17·82	60·24	4·33
1909	429·83	47·06	26·38	20·68	56·06	4·81
1910	439·05	51·14	27·16	24·00	53·10	5·46
1911	450·07	55·28	28·85	26·44	52·17	5·87
1912	465·15	61·65	30·16	31·49	48·92	6·77
1913/14	495·09	63·59	32·93	30·66	51·79	6·19
1914/15	519·22	60·43	32·75	27·68	54·19	5·33
1915/16	529·98	64·66	32·92	31·74	50·91	5·99
1916/17	535·28	70·68	33·40	37·28	47·26	6·96
1917/18	541·80	77·36	35·37	42·00	45·72	7·75
1918/19	549·74	86·29	41·80	44·49	48·45	8·09
1919/20	566·38	89·15	50·65	38·49	56·81	6·80
1920/21	626·80	91·99	60·29	31·70	65·54	5·06

Notes

(a) One crore = ten million. The rupee/sterling exchange rate was changing constantly during this period, mainly because the value of the silver-based rupee was declining against the gold-based pound. Conversion to sterling equivalents is therefore difficult. As an approximate guide, the value of the rupee was about 2s (£0·10) in 1870, 1s 6d (£0·07½) in 1890, and 1s 4d (£0·06½) in 1914.

(b) The figures for capital outlay are not altogether reliable. In particular, those for the years

1870-75, when the State began to construct railways on its own account, should be treated with some scepticism. As given here the figures represent revisions of those shown in earlier reports, though the basis of revision has not been stated. In addition there appear to have been changes in accounting methods, and in the definition of capital outlay. Figures in columns 1-4 have been rounded, so that the totals of columns 3 and 4 do not necessarily equal the sum shown in column 2.

Sources. *Railways in India*, annual administration report, P.P., 1909, LXIV, appendix No. 10; ibid., P.P. 1914/16, XLVIII, appendix No. 9; *Railways in India*, 1914/15 to 1920/21 (Simla: Government of India Railway Board, annual).

Notes

1 *Memorandum on some of the results of Indian administration during the past fifty years of British rule in India*, Parl. Papers, 1909, LXII [Cd 4956], p. 21.

2 See, e.g., Romesh Dutt, *The Economic History of India in the Victorian Age* (1906), passim; Daniel Thorner, 'Great Britain and the development of India's railways', *Journal of Economic History*, XI (1951). For a summary of some of the views that have been advanced on this issue see W. J. Macpherson, 'Economic development in India under the British Crown, 1858–1947', in A. J. Youngson (ed.), *Economic Development in the Long Run* (1972), pp. 144–5.

3 G. R. Hawke, *Railways and Economic Growth in England and Wales 1840-1870* (Oxford, 1970); R. W. Fogel, *Railroads and American Economic Growth: Essays in Econometric History* (Baltimore, 1964).

4 The few works dealing with the performance of the Indian railways have long been out of print. See especially Nalinaksha Sanyal, *Development of Indian Railways* (Calcutta, 1930), and R. D. Tiwari, *Railways in Modern India* (Bombay, 1941).

5 *Report of the Committee ... [on] the Administration and Working of Indian Railways* (Acworth Committee report), Parl. Papers, 1921, X [Cmd 1512], p. 12.

6 For a breakdown of the different types of administration, see *Report of the Committee on Indian Railway Finance and Administration* (Mackay Committee report), Parl. Papers, 1908, LXXV [Cd 4111], pp. 6–7.

7 Thomas Robertson, *Report on the Administration and Working of the Indian Railways*, Parl. Papers, 1903, XLVII [Cd 1713], p. 77.

8 G. A. Anderson, *Indian Railways: a Review of Mr. Robertson's Report* (Bombay, 1903), p. 33.

9 For a discussion of some of the drawbacks of the operating ratio as a measure of efficiency see R. J. Irving, *The North Eastern Railway Company, 1870–1914* (Leicester, 1976), apppendix I, pp. 286–7.

10 See D. H. Aldcroft, 'The efficiency and enterprise of British railways, 1870–1914', *Explorations in Entrepreneurial History*, V (1968); P. J. Cain, 'Railway combination and Government, 1900–1914', *Economic History Review*, second series, XXV (1972); R. J. Irving, 'The profitability and performance of British railways, 1870–1914', *Economic History Review*, second series, XXXI (1978).

11 See appendix, column 5. There were, of course, marked variations in the operating ratios of different lines belonging to the Indian system. That on the EIR was usually around 30 per cent, while on military lines, such as those of the North Western Railway, it was generally more than 100 per cent.

12 *Railways in India*, Parl. Papers, 1909, LXIV, pp. 3–4.

13 See, e.g., *Railways in India*, Parl. Papers, 1878, LVII, para. 64.

14 *Railways in India*, Parl. Papers, 1868, LI, para. 49 ff.; ibid., Parl. Papers, 1869, XLVII, para. 43 ff.

15 *Railways in India*, Parl. Papers, 1872, XLIV, appendix prepared by A. M. Rendel.

16 Average train load has been calculated by dividing goods train ton miles by goods train miles. Data derived from *Railways in India*, annual administration reports for 1881–84 and 1898–1902.

17 Juland Danvers, 'Defects of English railway statistics', *Journal of the Royal Statistical Society*, LI (1888), p. 13.

18 On leaving his position as Government Director of the Indian Railways in 1882 Danvers initiated a long campaign for the adoption of ton-mile statistics in Britain, which was subsequently taken up by other leading railway economists. See Danvers, 'Defects of English railway statistics', op. cit.; W. M. Acworth, 'English railway statistics', *Journal of the Royal Statistical Society*, LXV (1902); W. M. Acworth and G. Paish, 'British railways: their accounts and statistics', *Journal of the Royal Statistical Society*, LXXV (1911–12).

19 See E. H. Phelps Brown and S. J. Handfield-Jones, 'The climacteric of the 1890s: a study in the expanding economy', *Oxford Economic Papers*, new series, IV (1952), pp. 274–5; Aldcroft, 'The

efficiency and enterprise of British railways', *op. cit.* Note, however, the *caveat* on the inadequacy of overall productivity indices based on traffic movements to be found in Irving, *The North Eastern Railway*, *op. cit.*, pp. 76–7.

20 *Report of the Committee for the Revision of the Statistics of the Indian Railways* (Simla, 1880).

21 Acworth Committee report, *op. cit.*, P.P., 1921, X, p. 45.

22 Robertson Report, *op. cit.*, P.P., 1903, XLVII, p. 57. The quotation refers specifically to train speeds, but is clearly representative of Robertson's conclusions regarding railway performance in general.

23 Anderson, *Indian Railways*, *op. cit.*, *passim*, especially pp. 1, 34–7. Anderson was Secretary to the Government of Bombay Railway Department.

24 In the two decades before 1914 both the number of passenger journeys and the total ton miles carried per year more than trebled, total passenger journeys rising from 144·8 million in 1894 to 457·7 million in 1913/14, and ton miles from 4,862 million to 15,632 million in the same period. Figures from *Railways in India*, annual administration reports.

25 Robertson report, *op. cit.*, P.P., 1903, XLVII, p. 59.

26 Rai Sri Ram Bahadur, Proceedings of Council, 28 March 1906; *Financial Statement of the Government of India*, Parl. Papers, 1906, LXXXI, p. 179.

27 Acworth Committee report, *op. cit.*, P.P., 1921, X, p. 12.

28 Robertson report, *op. cit.*, P.P., 1903, XLVII, pp. 65–6.

29 Acworth Committee report, *op. cit.*, P.P., 1921, X, pp. 9 ff.; also Evidence, vol. III, paras. 3867 ff., 4412 ff., 5955 ff.

30 Robertson report, *op. cit.*, P.P., 1903, XLVII, pp. 80–2.

31 The underlying assumptions of Robertson's argument were criticised by Anderson, who questioned the purpose of having large loads, and of introducing high-capacity wagons 'merely because America uses them'. Thus, he maintained, the small consignments carried on many Indian lines would mean that the use of American-type bogies would not necessarily reduce, and might conceivably increase, the proportion of dead to paying weight. The difficulty with this criticism is that, though having some validity with respect to minor and non-commercial lines, major commercial lines such as the EIR would certainly have benefited from the use of larger rolling stock. Anderson, *Indian Railways*, *op. cit.*, p. 50.

32 Irving, 'The profitability and performance of British railways', *op. cit.*

33 *Ibid.*; also Cain, 'Railway combination and government', *op. cit.*

Acknowledgements

I am especially grateful to Mr P. J. Cain of the University of Birmingham, who supervised the original work carried out for this paper, and who made many helpful suggestions and criticisms. I would also like to acknowledge the useful comments made on an earlier draft of this paper by Professor D. H. Aldcroft and Dr C. J. Dewey of the University of Leicester.

8 The state and Indian railway performance, 1870–1920:

Part 2: The government, rating policy and capital funding

R. O. CHRISTENSEN

An analysis of some of the main criteria of efficiency of the Indian railway system in Part I of this paper showed that, though financial performance was good and improvements were made in traffic operation, there were, however, increasing problems, such as congestion and passenger overcrowding, from the end of the nineteenth century. These, it was argued, were the result mainly of the constraints imposed on the system by inadequate capital equipment at a time of rapid traffic growth.

Why did the Indian railways fail to develop so as to accommodate rising demand, and at the same time maintain, if not improve, standards of service? With regard to the interaction of supply and demand factors influencing the performance of the system, the Government of India occupied a key position. As the body ultimately responsible for the charges levied by the railway companies, and for the raising and allocation of funds for capital expenditure, it could determine to a considerable degree both the pattern of traffic development and the capacity of the railways to meet demand. In what respects, and to what extent, then, did government policy affect railway performance?

I

Rates and fares policy has been emphasised by several writers as a factor hindering the proper development of the Indian railway system.[1] The principal criticism has been that the rates were high, in relation both to operating costs and to the charges made in other countries, and discriminatory. In consequence they failed to encourage the optimum pattern of traffic growth, giving preference to external trade while ignoring the potential demand within India.

The rating policy of the railway companies was criticised on these grounds as early as the 1860s. The first guaranteed lines, based on the main ports of shipment, formulated their initial tariff policy with a view to securing the best earnings from the transportation of a small volume of high-class freight at high charges.[2] The government, however, concerned about the deficit on the guaranteed-interest account, took the view that the way to maximise revenues was to set the rates so as to provide the railways with the greatest possible volume of traffic. But though Juland Danvers expressed the hopeful opinion that on the question of charges 'the interests of both the Government and the Companies are the same',[3] there remained a considerable divergence between the respective policies of the two parties. The railway companies, each enjoying then a complete regional monopoly, maintained relatively high rates based on those prevailing in Britain, and the government's attempt to stimulate the growth of traffic by the imposition of maximum charges in 1868 proved largely ineffective. It was in response to this problem that A. M. Rendel undertook his analysis of costs as a means of discovering the possibilities of more efficient and economical working.

The improvements in loading which occurred in the 1870s led to some reductions in goods rates, although most were of telescopic rates for long-distance traffic, whereby charges fell as the distance carried increased. Critics of rating policy complained that this favoured import–export traffic at the expense of internal trade, but this form of discrimination was regarded by the railway companies as justifiable in view of the economies resulting from the handling of port traffic and carriage over long distances, and their position was subsequently upheld by the government.[4] It may be noted that in Britain similar complaints were made by farmers and other interest groups that they were being adversely affected by lower charges for imported produce conveyed from the ports, but a Board of Agriculture committee set up to examine the problem had little difficulty in accepting the arguments of the railway companies.[5]

There is, nevertheless, much evidence of concern on the part of the government that the tariff structure paid little attention to the development of internal goods and passenger traffic. The reductions which took place during the 1870s and 1880s demonstrated clearly the potential existing on this side of the traffic, and yet it was invariably left to the government to take the initiative. Thus in 1876–77 it prompted the companies to lower rates for grain in order to increase supplies to famine districts, with the result that goods receipts in 1877 were 35·7 per cent higher than they had been in the previous year.[6] The government was also instrumental in bringing about general reductions in lowest-class passenger fares, notably on the East Indian Railway (EIR) following its purchase by the State in 1879. Despite strong opposition from the operating company, third-class fares were reduced in 1882 from 3 to 2½ pies (or from about 0·34d to 0·29d) per mile, which led to an increase in passenger earnings on the line of over Rs 1 million (about £92,000) within three years.[7] Elasticity of demand was therefore high with respect to the lowest classes of both goods and passenger traffic, and the system appears to have had sufficient slack at this time to accommodate an increase in traffic resulting from lower charges. Generally low factor costs should also have favoured reductions, but for the most part the companies adhered to their existing policy.

Increasingly the government resorted to intervention through the setting of tariff limits, despite, or perhaps because of, the failure of the general maximum rates introduced in 1868. Anticipating the comparative approach adopted in the report made by Thomas Robertson in 1903,[8] the 1883 Financial Statement pointed out that, even though operating costs were lower in India than in the United States, rates were nevertheless higher. For example, the rate for food grains and minerals for the 954 miles from Delhi to Howrah worked out at 0·39d per ton mile, compared with 0·32d per ton mile for the 960 miles from Chicago to New York.[9] The Select Committee of 1884 also found that charges were still too high, and recommended that the government be given powers to change the maxima then in force.[10] Stringent control of company rating policies was established in 1887, when both maximum and minimum rates were stipulated, the latter on the grounds that under the guarantee system low charges not covering operating costs might be levied without harming the interests of shareholders, but causing a loss to the government.[11] The close regulation of charges did not prove successful, however, because the companies objected to the lack of discretion allowed them by an inflexible tariff structure in which the maximum and minimum rates for all classes of goods except the lowest were identical. Though a new schedule incorporating differentials was drawn up in 1891, interference by the government was resisted by many companies, and no further attempt was made to reduce general maxima, the revised structure remaining substantially unchanged until 1922.[12]

A further difficulty in the decades after 1870 was the absence of any uniform classification of goods or of standard rates for through traffic passing over more than one line. This drew charges that the railway companies were each acting as an *imperium in imperio*, and were using their monopoly powers to adopt 'individualistic' rating policies regardless of their joint interests or the requirements of traders.[13] The government argued that the various lines 'should, as far as possible, serve the country as if they were under one management',[14] and sought to secure a greater degree of uniformity, though with little more success than had been obtained by its efforts to reduce rates. A traffic conference held in 1884 to simplify charges fell through because of complaints by several companies, particularly the Great Indian Peninsula, that standardisation would lead to a loss of revenue. Measures to end tariff anomalies were adopted by the government in 1887,[15] but had subsequently to be modified after failing to win acceptance. Committees were set up in the first decade of the twentieth century to consider further the question of rate simplification, but the problem continued to be unresolved until 1915, when a system of uniform goods classification was finally accepted.

Much of the period under review, therefore, saw a clash on the question of rating policy between the government, which was concerned both to protect its own financial position and to represent the interests of railway users, and the companies. Relatively high charges, discrepancies between the rates set by individual lines, inter-company disputes arising out of encroachments on established spheres of influence, and the difficulties encountered by the government in its attempts to limit and regulate tariffs, all suggest little movement towards achieving a flexible and satisfactory structure of rates and fares. The period was not entirely one of immobility, however. From the late 1880s developments occurred which had some significant effects on charges and, consequently, on the pattern of traffic growth.

Perhaps the most important factor was the introduction of an element of competition between main-line companies. Regional monopolies began to be broken with the opening of the Rajputana–Malwa railway, which connected Bombay with the grain-producing districts of northern India. In 1884 the new line started to charge a low rate of 2·72 pies (about 0·31d) per ton mile for grain carried distances of over 400 miles, and in the following years the competing Bombay lines, the Great Indian Peninsula and the Bombay Baroda and Central India, gradually reduced their grain rates to the same level from 5·0 and 4·08 pies (about 0·57d and 0·47d) per ton mile respectively. By the mid-1890s similar reductions had been effected on all long-haul routes.[16] Coal traffic also benefited from increased competition. The opening of the Bengal–Nagpur railway in 1891 provided an alternative route to the EIR from Bengal to the western and north-western provinces, and the setting of low rates on this line did much to stimulate the internal trade in coal. The two decades from around 1890 were, indeed, characterised by rate skirmishes and undercutting on the part of trunk-line companies as monopoly gave way to monopolistic competition. Reductions continued until the first world war, when rising operating costs led the companies to enter into a cartel-type agreement ending competition between them.[17]

From the beginning of the twentieth century the government, too, played a part in bringing about a lowering of rates and fares, despite its previous limited success in effecting across-the-board reductions. In 1905 minimum rates for coal were reduced in order to help industries situated at a distance from the coalfields, and charges for long hauls subsequently fell by up to 50 per cent. Lowest-class passenger traffic showed a very marked growth as the result of certain tariff concessions. Thus in 1905 the EIR began to employ a telescopic fares system for third-class passengers, on a scale running from 2·50 pies (about 0·21d) per mile for the first 100 miles travelled to 1·50 pies (0·12d) for distances of over 300 miles. In the following year special return tickets were issued at one-and-a-half fare to facilitate the supply of plantation labour and attract pilgrim and *mela* (fair) traffic.[18] In response to measures such as these total passenger journeys approximately doubled during the first decade of the twentieth century.

Thus as the monopolies enjoyed by the main companies were partly eroded, and as the government used its powers to introduce cheap fares, the apparently high level of rates prevailing before the 1880s was greatly modified by important reductions for the largest-earning classes of traffic. Yet there were still those who believed that greater reductions were both possible and desirable. Robertson, for example, argued that, although Indian charges were low in comparison with those in most other countries (but not the US), they should have been even lower: 'the rates and fares in India should, broadly speaking, be only about one-sixth of those charged in England'.[19]

This argument was based on two considerations: first, the relatively low costs of construction and operation in India, and, second, that reductions on the scale proposed had taken place under pressure of competition in America. This line of reasoning was, however, flawed. Robertson himself had dealt at length with the growing inadequacy of Indian railway capacity to handle the traffic available. Therefore simply to point to low factor costs and the example of the American railways did not provide a very satisfactory basis on which to argue that the Indian rates were set too high. The observed high elasticity of demand, and the restricted provision of capital equipment, should have indicated the contrary, that by the beginning of the twentieth century a lowering of charges would only have exacerbated problems.

This is precisely what happened as a result of those reductions which were effected. So great was the increase in demand that within a few years the railway companies raised many charges in order to ease pressure on carrying capacity. Particularly after the outbreak of war in 1914, which led to a sharp decline in capital expenditure, the volume of traffic threatened a breakdown of the entire system. The government itself intervened in 1916–17, by raising the maxima for passenger fares, withdrawing concessions, and imposing surcharges on freight traffic, measures which reduced train mileage by nearly a quarter. But even had there been no war a reversal of the movement towards lower charges for staple freight items and lowest-class passengers would soon have been necessary. As the Acworth Committee pointed out, low rates and fares had substantially contributed to the problems of congestion and overcrowding, and one of its recommendations was that they should be further raised, above the higher levels introduced during the war, in order to lighten the burden created by excess demand.[20]

Ultimately, then, direct intervention by the government to lower rates, in conjunction with increased competition, itself encouraged by the government, contributed to the decline in standards of service which became evident in the twentieth century. From the 1860s, concerned that rates were too high in relation to potential demand, and with a view to obtaining the best revenues by attracting a large volume of traffic, the government used its powers to bring charges down. In addition, though organised pressure groups did not exercise the same influence as they did in Britain, it saw itself as the representative of the interests of traders and the travelling public. General economic objectives, the stimulus of commercial agriculture, the extension of the scope of trading activity, and the creation of a distribution network for food grains that would mitigate scarcities, were also a factor. But while there is much to support the view that before the 1880s the rating structure was highly conservative in relation to the potential for traffic growth, by the beginning of the twentieth century the balance had swung the other way, and the increase in traffic was outpacing the capacity of the railway system to handle it.

II

If the government played an important role on the demand side, it was perhaps even more directly involved on the supply side. Capital outlay on the first Indian railways was covered by a guarantee of a minimum rate of return, whereby shareholders in the companies were assured of at least 5 per cent on their investment even if the railways ran at a loss, the difference to be made up by the government.[21] When this system was abandoned in 1870 the responsibility for raising capital passed to the State, with significant consequences for subsequent railway development.

The guarantee had ensured a plentiful supply of capital; as Thorner has remarked, 'The capital which moved from England to India under these terms formed the largest single unit of international investment in the nineteenth century.'[22] The system soon gave rise, however, to charges that it had led to extravagant expenditure and poor performance. No line, with the exception of the EIR, was able to earn 5 per cent of its capital outlay, and by 1869 the government had advanced about £15 million to cover guaranteed-interest payments.[23] At the same time, estimates of the cost of construction proved to have been highly optimistic, as Danvers noted in 1868: 'Indian railways do not form an exception to the rule that expenditure always exceeds estimates. In some cases the cost has been three to four times greater than was expected.'[24] In government circles there was little doubt that the guarantee was to blame. Already in 1861 Sir Charles Wood, the Secretary of State, was urging that once the main trunk lines had been built no further guaranteed-interest contracts should be granted. An even stronger line was taken by Sir John Lawrence, the Governor General, eight years later:

> [It is clear] that the liability of the Government to a permanent and probably increasing charge on the revenues is much increased by the arrangement under which the Government can derive no profit whatever from the most successful railway, while it bears the entire cost of those which do not pay. There is no set-off of profit against loss in the Government share of these transactions. The whole profit goes to the companies, and the whole loss to the Government.[25]

Thus the guarantee system stood condemned. If the government was to pay 5 per cent interest on railway outlay, then it would be far better for the State to raise the capital through ordinary Government of India loan stock, and undertake railway construction and operation directly, than to continue to have to meet guaranteed-interest charges out of the revenues. It was therefore decided that from 1870 new lines were to be State-owned and operated, although existing contracts with guaranteed companies would remain in force until they expired.

The policy of funding new lines through government stock soon ran into difficulties. As early as 1876 a limit of £2½ million was placed on annual expenditure in order to prevent interest charges from becoming too great a burden on the revenues. (Not that this limit was observed in practice: in the later 1870s expenditure averaged about £4 million *per annum*.) Then, in 1881, the Secretary of State laid down that no works were to be constructed out of loans unless they were 'productive', that is, expected to defray the interest payable on capital outlay, with the exception of famine-protective works, which might be paid for directly out of the revenues. This policy was quickly reversed, however. Revenue difficulties at the end of the 1870s, arising out of famine and the second Afghan war, appeared to show that the government was not in a position to undertake railway investment entirely on its own account. At the same time the 1880 Famine Commission urged strongly that the policy of investing borrowed capital in railway extension should not be abandoned.[26] Consequently a mixed system of railway investment was introduced, whereby the government would be responsible for the construction of 'unproductive' or protective lines, leaving 'productive' lines to private enterprise. The entire outlay on 'unproductive' projects was expected to be spent in six years,[27] and it was clearly hoped that from then on all further expenditure would be undertaken by means of private capital.

The problem was that even lines classed as remunerative or 'productive' had little attraction for private capital in the absence of any guarantee. Only a very limited amount of capital was available within India, while in Britain hardly any interest was shown in providing the funds required for railway extension. The government's dissatisfaction was expressed in the Financial Statement for 1884:

> The Government of India is fully alive to the necessity and the advantages, in the interests no less of the State than of the public, of a rigorous policy in regard to its public works. But there are indications that the mercantile public, whose interest in the matter is little less than that of the State, while actively pressing for a policy such as that above indicated [i.e. for railway extension], and confident of prospective profits, is disposed to throw the burden and risks of such works as must be undertaken too entirely on the resources of the Government.[28]

With no improvement in the situation in the following decade, it became increasingly evident that sufficient private investment funds could not be raised unless they were again to be assured a high rate of interest. The government, indeed, reintroduced terms for the assistance of private capital in the 1890s, through the provision of rebates and then of a 3 per cent guarantee for branch line companies, but these measures met with little success. Nothing short of a return to the 5 per cent guarantee appeared to be adequate if funds on the scale required were to be attracted, but at a time when ordinary Government of India loan stock was paying 3 per cent there could clearly be no justification for such a move. The government, therefore, felt that it had no alternative but to raise the greater part of the capital itself.

State funding of railways, however, entailed financial arrangements that allowed the provision of neither a sufficient nor a consistent supply of capital. The railway account was subsumed in the government's general account, and was thus subject to the same budgetary rules and considerations as any other item. Although the limit set on government borrowing for railway purposes was occasionally raised, expenditure on the system was dependent to a very considerable extent on appropriations from central revenues, from provincial governments, and from the famine fund. The amount available from these sources took little account of the contribution made by net railway revenues, but constantly fluctuated in line with total revenues and government policy. In these circumstances the detailed planning of capital outlay was almost impossible. As early as 1884 the proposal was made that railway finances be separated from the general budget so as to allow the railway account to be maintained on the same principles as any other commercial enterprise.[29] But though the same suggestion was put forward in subsequent years, the Secretary of State continued to uphold the existing position.

Thus not only had the government to undertake almost all expenditure on the railway system on its own account, but it also had to do so within the limitations imposed by budgetary policy. In order to overcome much of the uncertainty and to make up for the deficient supply of private capital, in 1897 it inaugurated the so-called 'programme', a system whereby expenditure for a three-year period was sanctioned, the funds to be raised either directly by the government, or through companies working under contract to the government.[30] Capital funds for the first 'programme' were to be allocated mainly towards the construction of new lines, with the estimate for expenditure during the triennium 1896–97 originally set at Rs 28 crores (about £18·7 million), but subsequently extended to Rs 29·3 crores (£19·5 million).[31] This proved rather optimistic, however. In the next two years expenditure had to be revised downwards because of the demands of famine relief and military operations on the North West Frontier, and the total to March 1899 came to no more than Rs 25·2 crores (£16·8 million).[32] These oscillations established a pattern which was to be followed in succeeding years.

In order to avoid a recurrence of the embarrassments experienced in 1896–99, the government decided that the next triennial 'programme' should be more modest, and fixed a limit of Rs 20 crores (about £13·3 million) for the three following years. It explained its position thus:

> Having now carried to a fairly successful conclusion this policy of temporary activity in railway construction, we think it wise, both from a railway and from a financial point of view, to curtail for a time the rate of progress of railway construction. The nine or ten crores a year sanctioned in 1897 was not meant to be permanently adopted, and we deem it desirable now to allow a short time for the earning capacity of the lines recently constructed to develop itself, before again undertaking special burdens in the direction of capital outlay upon railways.[33]

But while this policy may have appeared to be prudent financially and expedient from the point of view of railway development, it bore little relation to actual practice. The sums allocated to the Railway Department under the 'programme' varied from year to year, in accordance with the state of the revenues, the British money market, and the general conditions, commercial or otherwise, prevailing in India. With the uncertainty regarding capital funding, and the claims made on the limited capital available by projects for extension on the one hand, and the need for improvements on the other, railway development was becoming increasingly caught in a conflict of priorities. Extensions of the system regarded as necessary by the government were being restricted because it was found impossible to keep down expenditure on open-line capital and projects already started to the amounts originally proposed. At the same time budgetary considerations, and particularly the demands made on the revenues by the 1899—1900 famine, meant not only that new projects had to be dropped but also that outlay on works in progress had to be reduced.

In order to resolve this dilemma the government established in 1900 a policy of giving first place to expenditure on improvements of existing lines, second to the completion of lines under construction, and third and last place to new projects. This arrangement, Robertson commented, would have worked well enough if the sums allotted to the railways had been sufficient; but they were not, and 'the actual effect is to restrict expenditure under all three heads'.[34] Difficulties were made worse by what came to be known as 'lapse', whereby unspent balances were suspended at the end of the financial year. Frequently the allotment made to one railway could not be spent, and in order to avoid a lapse in the total grant to the Railway Department the money had to be redistributed among other railways. As a result large grants were often made within a short period of the close of the financial year, and the work had then to be hurriedly undertaken. Conversely, budgetary exigencies often necessitated a sudden reduction of the grant, and work had then to be stopped in the middle of operations. The vagaries of the system of capital allocation were summarised by Robertson as follows:

> The uncertainty which always prevails as to the amount which will be allotted in any year, the doubt whether the amount available will really be available for expenditure, the sudden curtailment or withdrawal of grants, the equally sudden allotment of new and large grants at the close of the year which must be spent within a very short space of time, the effect this procedure has on the minds of Contractors in fixing their prices when framing their estimates — all this, I say, strikes at the very root of commercial success, and must lead to a waste of public money and to business being seriously hampered.[35]

The problems arising out of the shortage of capital were greatly exacerbated by the rapid growth of traffic in the early twentieth century. Despite the priority given to improvements on existing railways, even these lines were much behind in the provision of adequate facilities. Capital resources were becoming so stretched that the need for renewals of equipment, for additional rolling stock and for larger stations and goods depots could be only very partly satisfied, while the amount left for new construction dwindled to almost nothing.

The outlay necessary for railway development to keep pace with the rise in demand could not, therefore, be made under existing financial arrangements. Extraordinary revenue surpluses could not be relied on as a source of capital, and, despite a more liberal government loans policy, the raising of funds by borrowing was no easier than it had been in the 1880s. In the London capital market there was no lack of funds seeking investment outlets, but, it was explained, 'the number of investors to whom Indian stocks are familiar is relatively small'.[36] The government was reluctant to look for 'outside investors', however, on the grounds that the higher interest charges required would have adverse effects on the budget, and would also prejudice 'the good-will of our limited clientèle of habitual investors'.[37]

Robertson made two main recommendations with a view to enabling the Indian railways to be operated successfully along commercial lines. The first was that a semi-independent Railway Board be set up, which would have control of administration and working, including the construction of new lines and capital funding. This body would take over the functions of the Public Works Member of the Government, who, although the administrative head of the Railway Department, invariably had no training in railway management. Moreover the railways' subjection to the requirements of the Finance Department meant that they were 'reduced more or less to the same dead level as other Departments of Government'.[38] It was hoped that a Railway Board composed of railway specialists and having wide powers of decision-making would overcome these problems, but, though instituted in 1905, the Board proved to be less effective than had been envisaged. The Acworth Committee noted that its powers were actually very limited, and that the railways continued to occupy a subordinate administrative position as part of the Department of Public Works.[39]

Robertson's second recommendation was that a Railway Fund be established in order to ensure a steady rate of capital expenditure. The government would make an initial contribution of £10 million, and from then on the fund would be maintained from the surplus profits of the railway companies which accrued to the State. This proposal was not accepted by the Secretary of State, however, and expenditure continued to be made through annual allotments under the 'programme'.

A further attempt to tackle the problem of railway funding was made by the Mackay Committee in 1908, but this did not result in any radical reappraisal of the system of capital raising and allocation. Whatever the admitted shortcomings of the 'programme', the committee argued that it nevertheless 'appears to be the [procedure] best calculated to minimise the disadvantages of a restricted supply of capital'.[40] Nor, it held, was the question of 'lapse' as important as had been made out, since lapsed sums could always be re-allocated in the following financial year. The weakness in the argument was that in practice this was not the case. As the Acworth Committee subsequently pointed out, such sums 'may be re-allotted to the same railway, or may be re-allotted to another railway, or may not be re-allotted to railways at all'.[41] In general, the Mackay Committee was concerned not with the efficiency and development of the Indian railways as a large-scale commercial enterprise, but rather with how railway management 'could be fitted into the rigid framework of the financial system of the Government of India'.[42] The tenor of its report was therefore to suggest that the best would have to be made of the unsatisfactory arrangements in force.

The chief proposal of substance to be put forward by the Mackay Committee was that the government should fix a standard of annual capital expenditure so as to achieve a consistent rate of outlay. The standard recommended was £12·5 million (Rs 18·75 crores) per annum, but this was never attained before 1914, or indeed until after the first world war. As was to be expected from previous experience, the funds required to meet this standard were not forthcoming. The Mackay Committee assumed that the Secretary of State would be able to raise £9 million annually on the London capital market. When countries such as Argentina, covering a lesser land area than India and having a far smaller population, were borrowing larger sums, the supposition was superficially reasonable.[43] But Indian railway stocks were not considered an attractive investment. The average rate of return in the early twentieth century was never more than 3¼ per cent, while Latin American and North American railway stocks were paying between 4 and 5 per cent.[44] The government was not, however, prepared to attract a greater volume of funds by paying higher interest rates which would have had deleterious budgetary consequences.

The financial performance of the Indian railways far from justified this shortage of loan funds. Although the railway account was in deficit until 1900, this was almost entirely due to continuing interest payments made under the original guarantee, and to interest payments on outlay on lines still under construction, the burden of which was made heavier in certain years by the declining sterling exchange rate of the rupee. From the turn of the century, however, this deficit came to an end, and the railways became a source of increasing net revenues to the State. By commercial criteria, however, the measure of their financial performance is not the contribution made to the revenues but net earnings on investment. Ever since the trunk lines to the main ports had come into full operation in the mid-1870s the ratio of net receipts to capital outlay had never fallen below 4 per cent, and had usually been above 5 per cent.[45] Considered as a commercial enterprise, then, the Indian railway system was relatively profitable, and should have had little difficulty under ordinary circumstances in obtaining the funds required. The Acworth Committee, indeed, went so far as to argue that the government should have borrowed money at a rate higher than the net return earned by the railways.[46] Nevertheless, it was budgetary considerations rather than the inherent profitability of the railways which determined government investment policy. The railways, in consequence, had to continue to rely on the small number of investors in Indian stocks who were apparently willing to forgo the higher rates of return obtainable elsewhere.

Thus Indian railway investment became established in a pattern of stringency aggravated by uncertainty. Each year agents of the railway companies submitted forecasts for expenditure which, although higher than expectations, were well short of requirements. These forecasts were then amended by the Railway Board — 'the amendment is always in one direction, downward'[47] — and consolidated estimates of 'programme' expenditure were passed on to the Finance Department, which advised as to the amount likely to be available. Finally, the 'programme' was approved by the Secretary of State, although the grant remained provisional until the final grant was made a few days before the start of the financial year. Even this sum was liable to be altered according to the vicissitudes of the revenues and the demands made on them. The result was that expenditure on renewals and improvements was both insufficient and subject to discontinuities, while long-term projects were almost impossible to undertake.

The Acworth Committee likened the government to an industrial entrepreneur whose failure to invest in new machinery leads his firm to bankruptcy:

> With this difference: the manufacturer only brings down a single factory. The Indian case is that the railway undertakings, in which a great capital has been invested all over the country, have been held up for lack of the relatively small investment in new machinery required to make the whole of the plant efficient and economically productive. And there is another difference. If the single factory goes down, the customers can go elsewhere to fill their wants. The unfortunate customers of the Indian railways have nowhere else to go. They merely suffer. They are ceasing to suffer in silence.[48]

The most important of the recommendations made by the committee was that railway finance should undergo complete reform through the separation of the Railway Budget from the General Budget, and the freeing of railway management from the control of the Finance Department. Separate Railway Budget statements were published from 1924 onwards, but measures proposed for the reform of capital funding were given no full opportunity to show their effectiveness. Many of the Acworth Committee's recommendations were only half implemented, and then the railways were affected by further financial difficulties as the downturn of 1929 led to a decline in receipts and, consequently, retrenchment.[49] Thus the cumulative effects of half a century of capital shortage continued to be felt into the closing years of British rule.

III

It was argued in Part I of this paper that the performance of the Indian railway system in the period 1870–1920 was determined mainly by financial considerations. Financial criteria of efficiency did not, however, take sufficient account of those factors affecting real performance as shown by the ability of the system to develop in accordance with the growth of traffic. Nor was the concern with financial performance entirely consistent with allowing the railways to operate as a major commercial enterprise in their own right. The guiding principle of the government was that the railway system should be operated in the interest of the taxpayer, by meeting working expenses and charges on capital outlay, and over and above this providing a surplus for the General Budget, thereby reducing the need for revenue raised from other sources.[50] Government policy had, therefore, two main aspects: first, to maximise railway earnings by attracting as large a volume of traffic as possible, and, second, to ensure that capital investment did not have adverse effects on the revenues. The consequence of this policy was that from the end of the nineteenth century railway development faced increasing constraints, and standards of service suffered accordingly.

The State and Indian railway performance

The role of the State in influencing the pattern of railway development was itself determined by a combination of factors. In the 'free-trade imperialist' thesis, railways were to serve the interests of commercial and manufacturing capitalists by opening up India as a primary producer and market for British manufactures.[51] At the same time, however, they were also to serve the interests of the State by providing the means for internal control and external defence.[52] As a result of this conjunction of interests, the doctrine of economic *laissez-faire*, never perhaps applied as absolutely in practice as was prescribed in theory,[53] became especially modified in the case of Indian railway construction, through the provision of a guaranteed rate of return to shareholders. Yet when the guarantee was withdrawn, and once the most important trunk lines to the three major ports had been constructed, British private commercial groups lost interest in the further development of the Indian railway system. Railway investment in countries whose economies were growing more rapidly than India's, and which offered better rates of return than Indian railway stocks, proved more attractive.

The government could not so easily relinquish responsibility for the railways, however. There was a continuing military *rationale* for railway building; there was the need to construct lines for famine-protective purposes; and the demands of internal trade and the travelling public could not be ignored. The forces motivating the development of the system shifted entirely away from Britain to within the subcontinent itself. From 1870 onward the Indian railway system originally built and operated by 'private enterprise at public risk',[54] became essentially a State enterprise. Owned and operated by the State, or by companies operating under the auspices of the State, with capital either provided directly by the State or guaranteed by the State, the Indian railways were apparently exempted from the orthodoxy of *laissez-faire*. They were, however, certainly subject to another orthodoxy, that of the balanced budget. In consequence, the greatest economic enterprise undertaken by the British in India was consigned to a subordinate administrative position, and its development constricted by the provision of inadequate investment funds.

Notes

1 See, e.g., R. D. Tiwari, *Railways in Modern India* (Bombay, 1941), chapter 4; Daniel Thorner, 'Great Britain and the development of India's railways', *Journal of Economic History*, XI (1951), pp. 396—7.

2 *Railways in India*, P.P., 1864, XLIII, para. 36.

3 *Railways in India*, P.P., 1868, LI, para. 55.

4 P.W.D. No. 162 R.T., dated 2 March 1883, to the Government of Bombay; see Nalinaksha Sanyal, *Development of Indian Railways* (Calcutta, 1930), p. 180.

5 See P. J. Cain, 'Railways and price discrimination: the case of agriculture, 1880—1914', *Business History*, XVIII (1976).

6 *Railways in India*, P.P., 1878, LVII, para. 58.

7 *Railways in India*, P.P., 1884, LXIX, para. 97 ff.

8 Thomas Robertson, *Report on the Administration and Working of the Indian Railways*, P.P., 1903, XLVII [Cd 1713].

9 *Financial Statement of the Government of India*, P.P., 1883, L, para. 173.

10 *Report from the Select Committee on East India Railway Communication*, P.P., 1884, XI, para. 27.

11 Government of India P.W.D. Resolution No. 1446 R.T., dated 12 December 1887; in Supplement to the *Gazette of India*, No. 52, 24 December 1887. The fixing of minimum rates was unique to the Indian railways.

12 Goods rates maxima were raised in 1922 follow-

ing recommendations made by the Acworth Committee. For an account of rating policy to the first world war, see S. C. Ghose, *A Monograph on Indian Railway Rates* (Calcutta, 1918), especially chapters 1–2.
13 See Sanyal, *Development of Indian Railways, op. cit.*, p. 184; also Tiwari, *Railways in Modern India, op. cit.*, p. 162.
14 Letter by Sir B. Leslie, No. 633G, dated 25 August 1880, in P.W.D. Resolution No. 1446 R.T., dated 12 December 1887, *op. cit.*
15 See P.W.D. Resolution No. 1446 R.T., *ibid.*
16 See *Railways in India*, P.P., 1884, LXIX, p. 52; *ibid.*, P.P., 1886, XLIX, table p. 76; *ibid.*, P.P., 1893, LXIII, p. 148; *ibid.*, P.P., 1896, LXII, p. 168.
17 See Ghose, *Monograph on Indian Railway Rates, op. cit.*, chapter 2.
18 *Railways in India*, P.P., 1906, LXXXII, p. 6; *ibid.*, P.P., 1907, LIX, pp. 5–6.
19 Robertson report, *op. cit.*, P.P., 1903, XLVII, p. 69 *et seq.*
20 *Report of the Committee ... [on] the Administration and Working of Indian Railways* (Acworth Committee report), P.P., 1921, X [Cmd 1512], p. 32 and *passim.*
21 On the origins of the 5 per cent guarantee system see Daniel Thorner, *Investment in Empire: British Railway and Steam Shipping Enterprise in India, 1825–1849* (Philadelphia, 1950); also W. J. Macpherson, 'Investment in Indian railways, 1845–75', *Economic History Review*, second series, VIII (1955).
22 Thorner, *Investment in Empire, op. cit.*, p. viii.
23 Despatch from the Government of India, dated 11 March 1869, para. 11; *Papers Relating to Railway Extension in India*, P.P., 1868–69, XLVII.
24 *Railways in India*, P.P., 1868, LI, para. 39.
25 Minute dated 9 January 1869, para. 9; *Papers Relating to Railway Extension in India*, P.P., 1868–69, XLVII.
26 *Report of the Indian Famine Commission*, Part II, Measures of Protection and Prevention, P.P., 1880, LII, p. 171.
27 *Select Committee on East India Railway Communication*, P.P., 1884, XI, p. vi.
28 *Financial Statement*, P.P., 1884, LIX, para. 28.
29 S.C. on East India Railway Communication, P.P., 1884, *op. cit.*, Minutes of Evidence, Maj. Conway-Gordon, Under-Secy. to Govt. of India P.W.D., para. 3553.
30 A system providing for the funding of railway outlay over a three-year period had been in existence from around 1884, and was sometimes also referred to as a 'programme'. It was, however, somewhat *ad hoc*, and was much less extensive and systematic than the 'programme' introduced in 1897.
31 *Financial Statement*, P.P., 1897, LXIII, para. 74 ff.; *ibid.*, P.P., 1898, LXI, para. 58 ff.
32 *Financial Statement*, P.P., 1899, LXV, para. 22.
33 *Ibid.*, para. 26.
34 Robertson report, P.P., 1903, *op. cit.*, p. 31.
35 *Ibid.*, para. 91.
36 *Financial Statement*, P.P., 1902, LXX, para. 80.
37 *Loc. cit.*
38 Robertson report, *op. cit.*, p. 15.
39 Acworth Committee Report, P.P., 1921, *op. cit.*, pp. 36 ff.
40 *Report of the Committee on Indian Railway Finance and Administration* (Mackay Committee report), P.P., 1908, LXXV [Cd 4111], p. 10.
41 Acworth Committee report, *op. cit.*, p. 25.
42 *Loc. cit.*
43 Mackay Committee report, *op. cit.*, p. 14; also Acworth Committee report, *op. cit.*, p. 30.
44 See, e.g., R. J. Irving, 'British railway investment and innovation, 1900–1914', *Business History*, XIII (1971), p. 52.
45 See appendix, column 6, in Part I of this article.
46 Acworth Committee report, *op. cit.*, p. 27.
47 *Ibid.*, p. 23.
48 *Ibid.*, p. 27.
49 See *Report of the Indian Railway Enquiry Committee* (Delhi, 1937).
50 See, e.g., *Railways in India, Administration Report for 1920–21* (Simla: Government of India Railway Board, 1921), p. 42.
51 See John Gallagher and Ronald Robinson, 'The imperialism of free trade', *Economic History Review*, second series, VI (1953), pp. 4–5; also Peter Harnetty, *Imperialism and Free Trade: Lancashire and India in the Mid-nineteenth Century* (Manchester, 1972); Arthur W. Silver, *Manchester Men and Indian Cotton, 1847–1872* (Manchester, 1966). For a critique of the 'free trade imperialist' thesis as applied to India see C. J. Dewey, 'The end of the imperialism of free trade: the eclipse of the Lancashire lobby and the concession of fiscal autonomy to India', in C. J. Dewey and A. G. Hopkins (eds.), *The Imperial Impact: Studies in the Economic History of Africa and India* (1978).
52 See Macpherson, 'Investment in Indian railways', *op. cit.*
53 See Arthur J. Taylor, *Laissez-faire and State Intervention in Nineteenth-century Britain* (1972).
54 Thorner, *Investment in Empire, op. cit.*, chapter 7.

9 The German National Railway between the world wars: modernisation or preparation for war?

A. C. MIERZEJEWSKI

I

During the 1930s Josef Goebbels tirelessly presented to the world a picture of a National Socialist Reich that was powerful, monolithic and progressive. The overwhelming power of the Wehrmacht was of a piece with the victories of Mercedes and Auto-Union racing cars, the feats of German aircraft builders and aviators, and the successes of the German National Railway. The Deutsche Reichsbahn's (DR) leap to the apogee of European railroading seemed to begin as soon as the Nazis came to power. On 15 May 1933 the Fliegender Hamburger began its spectacular runs between Berlin and Hamburg at speeds up to 160 km per hour (99·2 m.p.h.). The successes continued and by 1939 the DR occupied the first thirty-two places on the European daily passenger train speed table.[1]

Yet behind the scenes there was uncertainty. As early as October 1937 the General Director of the Reichsbahn, Julius Dorpmüller, cautioned Hermann Göring, head of the Four-Year Plan, of the DR's need for more freight wagons and locomotives to support rearmament.[2] At the end of 1938 Dorpmüller warned Hitler directly that the Reichsbahn was at the limit of its potential.[3] On 23 June 1939 Colonel Rudolf Gercke, head of the transport section of the Army General Staff, informed the Reich Defence Council that the Reichsbahn was not ready for war.[4] Post-war apologists have eagerly adopted Gercke's view to relieve the Reichsbahn of responsibility for the defeat of 1945.[5] All analyse the inter-war history of the railway in the light of the lost war.

A shift in perspective from the narrow confines of military considerations to the broader perspective of the overall development of German transport since 1920 yields a more balanced picture. This article surveys the technological and organisational modernisation of the Reichsbahn from 1920 to 1939.[6] It addresses the question of whether the clear progress made by the DR after the

Nazi accession to power was the result of National Socialist war preparations or the product of long-term policies of the Reichsbahn itself.

II

When the Deutsche Reichsbahn was created on 1 April 1920, it faced massive problems stemming from Germany's defeat in the First World War.[7] The DR had to replace both the facilities and vehicles worn out during the war and the thousands of additional vehicles lost in the reparations settlement of 1919.[8] Given the weak finances of the young Weimar Republic, the Reichsbahn's leadership decided that it could master these problems only by becoming more efficient. It chose to cut personnel costs through technological modernisation and to rationalise its organisation.[9] There was a second external factor of ever-growing importance. For the first time Germany's railways faced serious competition from other means of transport. Motor vehicles challenged the railway's grip on passenger traffic and on high unit value, high profit margin, piece goods business. When Germany went to war in 1914, 55,276 passenger cars and 9,117 lorries travelled its roads. With the sale of surplus vehicles after the war, the number of lorries had jumped to over 30,000 by 1921. This tremendous growth continued and by 1931 there were 510,840 passenger cars and 161,072 lorries in the Reich, an eightfold increase in cars and a sixteenfold increase in lorries compared with 1913.[10] The 2–5 per cent share of freight ton kilometres carried by lorries is misleading. Because the lorries lured lucrative piece goods traffic away from the Reichsbahn, the loss of revenue that they inflicted on the Deutsche Reichsbahn-Gesellschaft (DRG) was disproportionately large. In 1929 the Reichsbahn calculated that it lost 220 million Reichsmarks (RM) in freight traffic and 190 million RM in passenger income to lorries, cars and buses.[11] In contrast, competition from aircraft remained insignificant. The inland waterways were viewed in most cases as welcome support in mastering the movement of low unit value, bulk commodities during periods of peak traffic. The exception here was the Mittelland Canal, which the Reichsbahn saw as a threat to its traffic between the Ruhr and Berlin.

The DR's leadership identified the new threat from the motor vehicle immediately and fought against it strenuously right up to the outbreak of the Second World War. It moved in two directions to counter the new challenger. On the one hand, the Reichsbahn sought to stifle its competition by trying to convince the government to create a transport monopoly that it would control. On the other, it introduced new high-speed passenger and freight services to woo customers back to the rails and hold on to those that it still had.[12]

The Weimar Republic passed laws in 1919, 1925 and 1931 to regulate motor vehicles.[13] The rationale behind the legislation was to eliminate competition that would damage the general welfare. Lorry owners were compelled to register and

Table 1. *Freight traffic in Germany between the wars*

Year	Total (ton km 10⁹)	Modal shares (%)		
		Reichsbahn	Water	Road
1924	60·1	78·8	19·1	1·9
1925	74·5	80·0	17·8	2·1
1926	81·3	79·7	17·9	2·2
1927	91·0	79·8	17·9	2·3
1928	91·0	80·4	16·7	2·8
1929	91·5	83·5	16·6	3·4
1930	79·2	77·0	18·6	4·4
1931	67·2	76·2	18·4	5·4
1932	58·8	75·5	18·5	5·9
1933	62·8	76·1	18·1	5·7
1934	74·4	76·5	18·1	5·4
1935	82·7	76·8	17·3	5·9
1936	93·7	75·4	18·2	6·3
1937	104·7	76·2	17·8	6·0
1938	125·7	79·8	14·9	6·7

Sources. Walther G. Hoffman, *Das Wachstum der deutschen Wirtschaft seit der Mitte des 19. Jahrhunderts* (Berlin, 1965), p. 417, Table 88; DR, *Geschäftsberichte*, 1924–38.

were to be allocated freight by the Reichsbahn. A system in which lorries and the railway complemented each other was to be established. Lorries would handle local traffic acting as feeders to the Reichsbahn's long-distance train services. These laws failed because they could not be enforced. Lorry owners sought business wherever they could get it, lorries operated by manufacturers and shop owners were not covered by the legislation, and the government lacked the manpower to monitor the industry effectively. The DRG took matters into its own hands by offering special discount rates (*Kraftwagentarife*) beginning in 1927.[14] In 1931 the Reichsbahn also completed an agreement with the Schenker forwarding company under which Schenker and associated freight expediters would collect piece goods, bring them to the railway stations and deliver them to their final destinations at the end of their railway journeys.[15] Yet, on the eve of Hitler's rise to power, the Reichsbahn sensed that it was losing its battle against the motor vehicle. Its attempts to create a transport monopoly had failed, and its

Table 2. *Passenger traffic in Germany between the wars*

Year	Total (passenger km 10^9)	Reichsbahn	Modal shares (%)			
			Streetcars Commuter railways	Buses	Autos	Aircraft
1926	59·3	72·3	25·4	1·5	0·7	0·02
1927	64·9	70·0	26·8	2·1	0·9	0·05
1928	69·1	68·9	26·3	3·5	1·3	0·04
1929	70·2	67·1	26·4	4·8	1·6	0·03
1930	64·8	66·8	26·5	4·5	2·1	0·03
1931	55·3	66·8	26·6	4·2	2·4	0·05
1932	46·6	66·1	26·6	4·7	2·5	0·06
1933	43·8	68·6	23·5	4·8	2·9	0·09
1934	49·8	69·9	21·9	4·4	3·7	0·06
1935	52·6	67·5	23·2	4·75	4·4	0·09
1936	61·9	70·2	20·3	4·5	4·8	0·19
1937	71·5	76·0	20·4	4·2	5·2	0·17
1938*	81·4	72·0	19·3	4·15	5·5	0·16

*Estimate.
Sources. Walther G. Hoffman, *Das Wachstum der deutschen Wirtschaft seit der Mitte des 19. Jahrhunderts* (Berlin, 1965), p. 417, Table 77, p. 692, Table 186; DR, *Geschäftsberichte*, 1926–1937.

rate discounts seemed to be hurting it more than the competition.[16] Frustrated from the outset with its attempts at regulation, and in spite of an unfavourable financial environment, the Reichsbahn had simultaneously remodelled itself along much more modern lines.

Fiscal problems greatly hindered the Reichsbahn in the implementation of its modernisation programme. The finances of the railways of the German states had been part of the state's budgets. That of the Reichsbahn had been part of the national budget when the railway was created. However, from 12 February 1924 the Reichsbahn's finances were separate from those of the Reich and, from October 1924 to July 1932, were tied to the reparations structure. Under the old regime, the railways had been administered in accordance with cameralistic fiscal principles that were aimed at producing revenue for the government and subsidising the overall development of the economy. In the new environment such a system was inadequate. So the railway adopted business principles of internal cost reckoning and profit calculation as supplements to the old methods.

Under the new system books were kept based on statistics depicting the financial affairs of each division (*Reichsbahndirektion*), repair shop, locomotive shed and yard, and for all materials consumed.[17] Time and motion studies were made of office and outdoor workers. In 1928 the Reichsbahn completed a very detailed study of the costs of every phase of its operations. It found, for example, that long-distance freight trains were its most profitable business and that first-class passenger services incurred heavy losses.[18] The avalanche of statistics was harnessed with Hollerith punched card machines, one of the first instances of their use in Germany.[19]

Even with the change in its official status, the Reichsbahn could not be managed like a private company. Circumstances beyond its control prevented it from pursuing a financial policy to facilitate the most rapid and thorough modernisation. As a quasi-state undertaking, the Reichsbahn was required to fulfil any demand for transport (*Beförderungspflicht*). It could not refuse traffic which would bring no profit and therefore lost heavily on service to thinly populated rural areas. Most importantly, the DRG's primary responsibility was to earn a sufficiently large surplus to meet its reparations obligations. The Allied agent-general for reparations and the Allied railway commissioner both had the power to influence the Reichsbahn's fiscal policies, and both looked askance at costly modernisation projects. In their view the Reischsbahn was already overcapitalised. Since the Allies were represented on the DRG's board of directors, they were well informed as to the Reichsbahn's plans, although they lacked sufficient votes to block them in that venue. The German members of the board, nevertheless, had every incentive to avoid alienating their Allied counterparts and especially to avoid default. Under the terms of the Dawes Plan, the Allied railway commissioner could seize control of the Reichsbahn should it fail to pay its reparations bills. Both German public opinion and the Weimar government sought to avoid such a catastrophe at all costs. Therefore the Reichsbahn's board of directors, led by the fiscal conservative Werner von Siemens, followed a fairly restrained credit policy. This ensured that the DRG always paid reparations according to schedule, but it also caused the German rolling stock industry considerable hardship due to lack of orders and slowed modernisation.

The domestic credit market was extremely tight in Germany throughout the 1920s and 1930s. The Reichsbahn found it very difficult to compete for credit against private concerns and other government institutions such as the post office. A variety of schemes was attempted to obtain credit to enable the Reichsbahn to modernise. Some involved loans from the railway industry itself, such as from the German Carriage Manufacturers Association or from Siemens AG.[20] Others involved subsidies from municipalities and states, as in the case of the loans that made possible the construction of the new main station in Stuttgart.[21] Another illustrative case occurred in 1928. The Reichsbahn sought money to purchase new wagons. A group of American banks led by Robert

Dillon & Co. offered the DRG 'equipment bonds'. Under this procedure, frequently employed in the USA, the bankers would purchase vehicles from a group of manufacturers. The bank consortium would sell bonds to cover the cost of the order. The vehicles would then be rented to the railway. They would become the property of the railway only when it had paid the full cost of the purchase. After a thorough investigation of such instruments, the Reichsbahn's directors chose to pursue the opportunity, so long as it could be transacted through a German bank. Negotiations proceeded with the Dresdener Bank into May 1928 and seemed to be approaching a successful conclusion when the agent-general for reparations, Parker Gilbert, intervened. Gilbert feared that the sale of equipment bonds would siphon money from the international credit market and thereby hinder German industry from making its reparations payments. He also objected on the ground that he had not been informed in advance.[22] By the end of May both the Reich Finance Ministry and the Allied railway commissioner, Gaston Leverve, had also moved to block the transaction.[23] Finally, in July, Dorpmüller capitulated and informed Gilbert that the Reichsbahn had abandoned its intention of accepting the offer of equipment bonds.[24]

Table 3. *Reichsbahn real capital growth 1925–34 (million RM)*

Year	Total	Internally financed	Externally financed
1925	480·4	241·4	239·0
1926	421·5	13·9	407·6
1927	524·9	176·4	348·5
1928	302·1	16·4	285·7
1929	203·6	17·2	186·4
1930	222·1	13·1	209·0
1931	77·8	12·9	64·9
1932	50·2	4·4	45·8
1933	165·1	15·2	149·9
1934	252·9	17·3	235·6
Total	2,700·6	528·2	2,172·4

Source. Deutsche Reichsbahn-Gesellschaft, *Statistische Angaben über die Deutsche Reichsbahn im Geschäftsjahr 1934* (Berlin, 1935), p. 87, 'Übersicht über den Anlagezuwachs seit 1925 und seine Finanzierung'.

Despite these difficulties, the Reichsbahn made a considerable net investment in modernisation between 1925 and 1934. Most real capital growth was supported by external loans and the sale of stock. Of the external capital, 581 million RM were obtained through the sale of preferred stock, 178·1 million through write-offs, 151·2 million from construction loans from states and communities, and 1,252 million through new loans and bills of exchange.[25] The external obligations incurred for modernisation constituted 79 per cent of the overall debt of the Reichsbahn. Significantly, net investment by the DRG constituted only 3·8 per cent of total German net investment from 1925 to 1929, a smaller share than during any five-year period since 1851.[26]

Its financial room to manoeuvre restricted, the Reichsbahn sought to free liquid assets by saving money through structural efficiencies. Possibly the single most important initiative was the very creation of the Reichsbahn. Before 1 April 1920 there was no German National Railway, but seven state railways.[27] Although particularism was a stubborn obstacle, with Bavaria receiving special concessions, unification made possible the standardisation of procedures and equipment and the gradual elimination of redundant organisations. The Magdeburg division was dissolved entirely, and the Würzburg and Nürnberg divisions were merged.[28] By 1929 the DRG had closed twenty-two main repair shops (*Reichsbahnausbesserungswerke*) and reduced sixty-one others in size.[29] Those repair shops that remained adopted production-line methods, concentrated on fewer types of vehicles each, standardised their organisations, were grouped under regional directorates (*Geschäftsführende Direktionen*) and introduced piece-work rates as incentives to the workers.[30] In 1927 alone the Reichsbahn eliminated over 5,000 jobs through such measures.[31] Similar administrative steps were taken throughout the railway. Between 1923 and 1930, for example, 336 offices and sixty-eight stations were closed.[32]

In addition, the Reichsbahn reformed its management of material stocks during the 1920s. It eliminated holdings of many items and closed some warehouses entirely. It conducted a census of stocks, re-allocated them according to real needs and reduced them in scale whenever possible. It substituted better, but less costly, materials and introduced a semi-automated reporting and ordering system to manage them more efficiently.[33]

The Reichsbahn made a major effort to accelerate traffic through both managerial and technical initiatives. Rolling stock usage was rationalised, based on statistical studies, and placed under stricter control. The 'dispatching system' was adopted from American practice, and special offices (*Zugüberwachungsstellen*) were created to expedite movements on especially troublesome stretches and in difficult yards.[34]

Among the most important technical improvements were those made to the Reichsbahn's steam locomotive fleet. When it was formed in 1920, the DR inherited over 210 different types of steam engine from its predecessors.[35] The problems of stocking spares for such a variety were insuperable. The need to replace thousands of locomotives provided the Reichsbahn with the opportunity to prune out the less valuable types and to standardise new models. From 1925 to 1927 alone the DR retired 5,3000 steam locomotives.[36] In 1923 it implemented the so-called *Einheitsprogramm* (Standard Programme) concentrating on just fourteen types of new engine.[37] The Reichsbahn selected conventional, though highly refined, configurations featuring two or three cylinders while other railways were opting for four.[38] It experimented with turbine, pulverised coal and high-pressure locomotives but did not adopt them because they promised no real cost advantages.[39]

The results of the programme were none the less impressive. By 1930 the average Reichsbahn steam locomotive was 39 per cent more powerful than its 1913 predecessor.[40] Its actual pulling power was up by about 50 per cent.[41] Its own weight had declined by 3 per cent, its fuel consumption by 19 per cent and its demand for crew by 18 per cent.[42] The number of different locomotive types had fallen by 38 per cent, due largely to retirements.[43] From 1920 to 1929 the railway obtained 6,241 new locomotives, with most being delivered during the first four years. The annual average for orders of new steam engines from 1924 to 1929 of 113·8 was more representative.[44] The decline in locomotive orders beginning in 1924 had two causes. One was the surplus of locomotives;[45] the other was lack of money. The Reichsbahn spent a total of 105·5 million RM on new steam locomotives in the latter period, or about 17·6 million RM per year.[46]

During the 1920s the DRG engaged in only limited expansion of its electrified system because of the high initial costs involved. It concentrated on areas that offered easy, cheap access to sources of electricity and where the advantages of electric traction could be realised on mountainous lines.[47] By 1928 the Reichsbahn had increased its electrified network by almost three times compared with 1920. However, that still represented only 2·3 per cent of its entire system. In the same period the DRG expanded its fleet of electric locomotives by 2·5 times to 398.[48] As with steam locomotives, the Reichsbahn rationalised the purchase of electric engines to seven basic types, with an initial order for 138 issued in 1922.[49] All incorporated advanced motors and suspensions that allowed them to outperform comparable steam locomotives. However, the prohibitive costs of electrifying lines caused the DRG to opt instead for the diesel railcar as its answer to automobile and bus competition.

The diesel railcar offered the Reichsbahn the possibility of greatly reducing operating costs while enhancing service. During the early 1920s, the rationale for building diesel railcars was to satisfy low density rural demand. By the end of the decade it had evolved to providing high-speed flyers connecting major cities. In both cases the diesel railcar offered advantages of low weight and modest cost per passenger. Significantly, these advantages could be gained on existing lines

Table 4. *Locomotives ordered by the Reichsbahn, 1919–34*

Year	Number	Year	Number
1913	1,690	1927	307
1919	2,730	1928	101
1920	1,157	1929	26
1921	1,772	1930	76
1922	1,206	1931	115
1923	1,423	1932	106
1924	50	1933	146
1925	30	1934	220
1926	169	–	–

Sources. DRG, Hv, 'Bestand, Durchschnittsalter, Ausmusterung und Beschaffung von Dampflokomotiven, Personen-, Gepäck- und Güterwagen der Deutschen Reichsbahn während der letzten 10 Jahren', 30.73a Fuv 100, 2 November 1935, Anlage 4, BA R5/2541.

without a heavy investment to add wires, towers and transformer stations.[50]

The Reichsbahn introduced diesel railcars on a variety of local runs during the mid-1920s.[51] At the same time the Society for Transport Technology, formed in 1924, began conceptual work on a lightweight railcar that would operate at speeds of 200–300 km per hour (120–80 m.p.h.). The upshot was the propeller-driven Schienenzeppelin (Rail Zeppelin) designed by Franz Kruckenberg and his Flugbahn-Gesellschaft. Safety problems, as well as low passenger capacity, convinced the DRG not to pursue this line of development.[52] Instead, it opted for a more conventional diesel railcar designed to operate at speeds of 150–60 km per hour (93–9 m.p.h.).[53] In 1931 the Reichsbahn contracted with the Waggon und Maschinenbau AG of Görlitz to build a two-car articulated set to accommodate 102 passengers using two, twelve-cylinder, 410 h.p. Maybach diesel engines. The form of the vehicle was determined in the wind tunnel at the Zeppelin factory at Friedrichshafen.[54] The result was a smoothly shaped two-car set that was 41·9 m (137·5 ft) long and offered 102 passengers comparatively spartan accommodation. On 15 May 1933 the train, dubbed by the public 'Fliegender Hamburger', made its first scheduled run on the relatively flat and straight line from Berlin to Hamburg at the startling average speed of 124·7 km per hour (77·3 m.p.h.). The DR now had the fastest train in the world.[55]

In addition to its diesel railcars, the Reichsbahn also improved its regular passenger carriages and goods wagons. It concentrated on just three types of carriage for main lines, and one for secondary lines. All used welded steel construction.[56] Express train carriages received new suspensions cleared for running up to 150 km per hour (93 m.p.h.).[57] By 1933 18·2 per cent of all carriages, and 29·3 per cent of all express carriages, conformed to the new standards.[58] The most important innovation in freight was the introduction of wagons, for carrying bulk commodities, with special automatic unloading features, some of which were capable of payloads of up to 60 tons. The DRG also completed its predecessors' programme of equipping all goods wagons with Kunze-Knorr air brakes. This programme could be completed only with the help of a loan from the Kunze-Knorr company.[59] In 1930 the Reichsbahn calculated that it had reduced its goods train crews by 47 per cent (28,000) by the introduction of air brakes alone.[60] Meanwhile, special magnetic brakes were perfected to slow high-speed trains. The Fliegender Hamburger, among others, was equipped with them.[61] Safety improved with the introduction of Inductive Train Control (Induktive Zugsicherung-Indusi) which prevented a train from overrunning a stop or slow signal.[62] Indusi consisted of an electromagnet on the locomotive that triggered safety mechanisms on the engine when passing signals. If the engineer did not respond either by pressing a button to show that he was conscious, or by slowing the train as appropriate, the locomotive would be stopped by Indusi. Overall, the DRG spent 1,090 million RM on new vehicles of all kinds from 1925 to 1932, an average of 135·8 million RM annually.[63]

If the Reichsbahn had improved only its vehicles, it would have failed to become more efficient and more competitive. About two-thirds of its income came from freight traffic, and the decisive factor in freight operations was marshalling. So the DRG focused a great deal of effort on enhancing the performance of its marshalling yards. One of the most effective measures here was the introduction of small diesel shunting locomotives. These little machines of no more than 65 h.p. relieved the main line locomotives of shunting in small stations, thereby saving time and money. The Reichsbahn estimated that these *Kleinloks* cut marshalling costs by one-third to one-half.[64] Marshalling facilities themselves were mechanised to the greatest extent possible, including the installation of retarders and remotely controlled points. Humps were rebuilt after careful definition of their optimum shape, points relocated, new administrative procedures introduced and radios installed in some yard locomotives. Fourteen major yards and many small ones were renovated.[65] By 1929 trains that formerly took twenty to thirty minutes to break up could be handled in six to seven minutes in Ruhr yards like Hamm and Osterfeld Süd.[66]

At the same time, the DRG improved other components of its fixed plant. New passenger stations were built, as in Stuttgart. Centrally operated, electrically actuated points and signals were installed in selected stations, and the signals

Table 5. *Reichsbahn vehicle expenditure (million RM)*

Year	Vehicles	Locomotives	Year	Vehicles	Locomotives
1919	–	66·6	1929	203·7	6·2
1920	–	178·8	1930	172·8	11·9
1921	–	120·8	1931	103·7	18·2
1922	–	102·6	1932	69·6	20·6
1923	–	103·3	1933	88·4	23·9
1924	–	12·3	1934	130·0	22·6
1925	75·7	13·7	1935	130·0	–
1926	64·1	17·2	1936	125·0	–
1927	202·0	24·9	1937	138·0	–
1928	212·4	31·2	1938	204·0	–

Sources. DRG, Hv, 'Bestand, Durchschnittsalter, Ausmusterung und Beschaffung von Dampflokomotiven, Personen-, Gepäck-und Güterwagen der Deutschen Reichsbahn während der letzten 10 Jahren', 30.73a Fuv 100, 2 November 1935, Anlage 7, BA R5/ 2541; Ursula-Maria Ruser, *Die Reichsbahn als Reparationsobjekt* (Freiburg, 1981), p. 356, Anlage 13, also available in BA R43 II/187b; Deutsche Reichsbahn, *Geschäftsberichte*, 1933–38.

themselves were redesigned to be easier to read.[67] Sections of track began to be replaced on a regular schedule for the first time.[68]

The Reischsbahn exploited its innovations to offer the public new services. In 1924 it introduced unit trains for coal transport composed of high-capacity wagons linking the Ruhr, Berlin and Upper Silesia.[69] At the opposite end of the spectrum, special high-speed, light goods trains, many using diesel railcars, were run to accelerate express freight shipments.[70] Goods train speeds rose as a result of these improvements. For example, by 1925 the running time between Berlin and Cologne had been cut to fifteen hours compared with twenty-nine in 1913.[71] Passenger traffic was also speeded up. On 1 June 1923 the Reichsbahn began running long-distance express trains, the Fernschnellzüge (FD-Züge).[72] They contributed to a rise in the average speed of all express trains of 24·3 per cent by 1932.[73]

The net result of these initiatives was a major improvement in the Reichsbahn's operating efficiency. For example, the average goods train carried one-third more freight in 1929 than its predecessor of 1914.[74] In 1929 each goods wagon carried 18·2 per cent more freight than in 1921. The ratio of train kilometres per employee improved by 18·7 per cent compared with 1926, though this was still 3·8 per cent below the performance of 1913.[75] In 1929 the DRG earnt 26·3 per cent more per passenger, and 65·4 per cent more per ton of freight than in 1913.[76] Income per employee was 70 per cent higher.[77] Yet the Reichsbahn still perceived that it was losing the transport race to cars and lorries.

III

The Reichsbahn sought two things from the Nazi government: financial assistance and support against the motor vehicle. At the board meeting on 29 May 1934 Dorpmüller said that his main purpose was to maintain technological progress to make the Reichsbahn more competitive. At the same time Max Leibbrand, head of the DRG's Operations Section, said that the Reichsbahn's aim was to defeat the cars and lorries.[78] Although Dorpmüller and his colleagues welcomed the Hitler regime, they were quickly disappointed by the Führer's attitudes towards transport. As early as 10 April 1933 Hitler informed the General Director of the Reichsbahn that the railway would be subordinate to road traffic.[79] Nevertheless, the DRG pressed forward with its existing modernisation policies.

The new regime's emasculation of the state governments allowed the Reichsbahn to centralise its administrative apparatus more fully. On 1 September 1933, as part of the regime's drive against particularism, Bavaria's special railway status was eliminated.[80] On 1 April 1934 the Nazi regime annulled the fourteen-year-old agreement between the Reich and the states that had created the DR.[81] This smoothed the way for the elimination of two divisions: RBD Oldenburg, on 3 July 1934; and RBD Ludwigshafen, on 31 March 1937. Their regions were re-allocated to their neighbours RBD Münster, on the one hand, and RBDs Saarbrücken and Mainz, on the other.

Technical advances also continued. On 29 May 1934 the Reichsbahn completed its plan to expand service using diesel railcars.[82] The DRG wanted to buy 4,271 sets of four basic types to take over about 60 per cent of its passenger traffic.[83] It estimated that 1,500 million RM would be needed to complete the programme, of which 50 million RM had already been spent or budgeted.[84] Not surprisingly, given state policy, the Reichsbahn did not receive financial support from the government. Instead, it implemented its expansion plan at a slow pace by using its own resources and postponing completion by twelve to seventeen years. Very few of the railcars intended for local traffic were procured, and only thirty-four additional high-speed sets, two of which were electric, were obtained before the war.[85] Yet these few vehicles allowed the DRG to establish a system of nine routes covering 6,000 km (3,720 miles) that served eight major cities as part of the FD network.[86]

Purchases of steam locomotives continued at a steady pace. From 1933 to 1938 the Reichsbahn ordered an average of 144 locomotives per year.[87] This was 27·4 per cent greater than during the period 1925–29, but hardly sufficient to match the concurrent growth in traffic. A few specialised speedsters made their appearances. In 1935 the fast Henschel–Wegmann train began operation between Berlin and Dresden. But it was slower than the Fliegender Hamburger and had no sequel.[88] Borsig built the series 05 that seized the world speed record for steam locomotives with a run of 200·4 km per hour (124·2 m.p.h.) on 11 May 1936. Only two were completed.[89] The Reichsbahn developed two new electric locomotives but did not produce them in significant numbers either.[90] The overall increase in locomotive power continued but slowed to 3·7 per cent between 1935 and 1938.[91] Goods train transit times fell by about one-third between major centres.[92] There was a thirteenfold increase in small shunting locomotives.[93] The FD service was expanded to form a system linking Berlin with the major economic centres of the Reich. Express train speeds continued to rise, increasing by 23·3 per cent beween 1932 and 1939, or at about the same rate as during the Weimar era.[94] In 1935 the DRG began the acquisition of a new class of express coaches.[95] Revealingly, though, unlike its predecessors, the Reichsbahn drew up no plans to build war locomotives should the need arise.[96]

Table 6. *Reichsbahn finances 1924–39 (million RM)*

Year	Income	Expenditure	Debt
1924	–	–	585
1925	4,669.1	3,975	624
1926	4,540·8	3,681	949·7
1927	5,039·3	4,159	1,006·7
1928	5,159·2	4,294	1,247·5
1929	5,353·8	4,493	1,256·4
1930	4,570·3	4,090	1,676·9
1931	3,848·7	3,622	1,712·7
1932	2,934·3	3,001	1,920·3
1933	2,920·6	3,057	2,332·2
1934	3,326·3	3,302	2,741·8
1935	3,586·1	3,434	2,770·6
1936	3,984·1	3,513	2,645·4
1937	4,42	4,124	2,371·2
1938	5,13	4,88	2,791·9
1939	5,813	5,465	3,489·4

Sources. Deutsche Reichsbahn-Gesellschaft, *Statistische Angaben über die Deutsche Reichsbahn im Geschäftsjahr 1934* (Berlin, 1935), p. 318, Bild 3; Deutsche Reichsbahn, *Geschäftsberichte*, 1935–39; Deutsche Reichsbahn, 'Schuldenstand der DRB seit 1924', 31 December 1943, BA R5/2549.

In short, the DRG pursued its existing policies aimed at motor vehicles as if no war were contemplated by the Nazi regime.

Changes occurred in construction of fixed plant. Signals, rails and points continued to be improved. Noteworthy was the introduction of the Railway Automatic Connecting Equipment (Bahnselbstanschlussanlage-Basa) that greatly enhanced the railway's internal telephone service in 1933 and 1934.[97] The location of signals was revised to accommodate the higher speeds achieved by the FD- and Fliegende Züge.[98] Some new facilities were built, such as the passenger station at Stettin and the locomotive sheds at Breslau. Other facilities were modernised, such as the important marshalling yard at Gleiwitz in the heart of the Upper Silesian coal region.[99] Simultaneously, vast resources were squandered on Hitler's grandiose schemes to remodel the cities of Berlin, Munich, Hamburg and, later, Linz. These massive schemes involved moving entire stations across cities which diverted engineers, materiel and money from more important projects.[100]

As for the Reichsbahn's finances, subsidies from the government were not forthcoming. In the autumn of 1935 both Dorpmüller and the Transport Minister, Peter Paul von Eltz-Rübenach, appealed to the regime for fiscal relief. Although reparations payments had ceased three years earlier, the DRG was compelled to pass the proceeds of the transport tax to the government as it had under the reparations system.[101] Dorpmüller contended that although the railway's income was little changed from 1913, it had paid 1,330 million RM in job creation bills since 1931, 100 million RM in subsidies to national defence projects in the form of rate discounts and 490 million RM in other 'political expenses' imposed by the government.[102] The response of the Finance Minister, Schwerin von Krosigk, was extremely negative. He told the DRG to balance its budget on its own, preferably with a rate increase. He counselled the DRG not to be concerned about the decline in traffic and to cut its internal expenses.[103] In 1936 Hitler did not include the DRG in the Four-Year Plan, thereby making it difficult for the Reichsbahn to obtain trained personnel, raw materials and factory space.[104] Net investment by the Reichsbahn represented only 1·7 per cent of overall Reich net investment between 1935 and 1938. This was a new low in the history of German railways.[105] Yet the Reichsbahn's overall expenditure on vehicles averaged 145·4 million RM from 1934 to 1938, somewhat higher than during the 1920s.[106] The DRG was able to sustain its capital spending and pursue the new projects assigned to it by the Hitler regime only by going heavily into debt. In short, Reichsbahn capital spending remained remarkably stable while market conditions changed drastically.[107]

Traffic volume rose dramatically and, especially after 1936, new traffic patterns emerged due to the dispersal of industry occasioned by rearmament. Simultaneously, competition sharpened due to the motorisation policies of the Nazi regime. Indicative of these changes, in 1938 the DR moved 23·7 per cent more passengers, and 37 per cent more freight than in 1928. Average haul for freight rose by 19·2 per cent. The average length of a passenger trip rose by about 13 per cent.[108] In November 1933 the Reichsbahn estimated that although only one-twelfth of the volume of its freight traffic was threatened by long-distance lorries, that segment represented fully one-third of its freight income.[109] In 1935 the Finance Section of the Reichsbahn noted that traffic was not rising as quickly as economic activity. It attributed the shortfall to road competition. The Traffic Section estimated that in 1935 the DR lost 350 million RM in passenger revenues and 300–20 million RM in freight revenues to motor vehicles.[110]

Aside from modernising its services, the DRG continued to attempt to meet the challenge of the motor vehicle by rate cuts and regulation. In 1934 it introduced major passenger fare reductions.[111] On 26 June 1935 a law on Long-Distance Freight Traffic–Motor Vehicle Traffic was promulgated by the government. Freight hauled by lorries beyond a radius of 50 km from point of origin was strictly controlled. All lorry companies were compelled to join the Reich Lorry Operators Association (Reichskraftwagen-Betriebsverband — RKB).[112] The legislation did not have the desired effect for the same reasons that earlier efforts at regulation had failed. Consequently, Dorpmüller, now Transport Minister, issued a series of orders in 1938 to rein in the competition. Again, the allocation of freight was to be made by the RKB and the Reichsbahn.[113] Yet piece goods traffic contined to shift to the lorries.

Inroads into the Reichsbahn's share of passenger traffic were, if anything, even more severe. Bus traffic in 1924 generated only 500 million passenger kilometres. By 1929 it had jumped to 3,400 million, or about 7 per cent of the Reichsbahn's total. Near the end of the depression, in 1933, it remained a healthy 2,100 million passenger kilometres, still 7 per cent of the Reichsbahn's performance.[114] By 1938 about one-quarter of all long-distance passenger traffic moved by road.[115]

Ironically, the Reichsbahn was forced to subsidise the autobahns. When the Reichsautobahen Gesellschaft was formed on 27 July 1933, it was made a subsidiary of the Reichsbahn. Rather than facilitating control by the DRG, this opened the door for the exploitation of the railway for the benefit of the highways. The DRG gave the new company 50 million RM in start-up capital and provided it with administrators and technicians. Reichsautobahnen freight was carried at the same low rates as the Reichsbahn's own service goods. In 1935 and 1936 the DRG gave the highway company a total of 800 million RM from the 1,000 million it had borrowed from the government.[116] On 1 July 1938 responsibility for the Reichsautobahnen was removed from the DR. Yet the Reichsbahn was still required to pay its competitor an annual subsidy of 50 million RM.[117]

The Reichsbahn reacted by doggedly pursuing the policies that it had initiated during the Weimar era. Since Dorpmüller was not part of Hitler's inner circle, he could not anticipate how Hitler's expansionist intentions would generate a greater demand for transport and lead to war. From 1933 to 1936 the DRG perceived no need for additional rolling stock.[118] Later, when demand seemed to approach the Reichsbahn's limits, Dorpmüller and his staff responded by preparing a series of schemes for increased vehicle production that would realise the DRG's vision of modernisation. The first was completed in mid-December 1937. It proposed the acquisition of 117,790 new goods wagons, leading to a real growth of 82,630.[119] By the end of December this programme was expanded to include locomotives and fast diesel railcars.[120] On 13 January 1938 Dorpmüller ordered the preparation of a supplement to this programme including small shunting locomotives and diesel railcars.[121] Dorpmüller's state secretary, Wilhelm Kleinmann, ordered a re-casting of this plan in July 1938 to take into account Germany's recent territorial acquisitions and postponed its implementation by a year. Again, the latest types of rolling stock intended to counter motor vehicle competition were included.[122] The strains resulting from rearmament compelled Göring to accept the DR's massive vehicle building programme on 14 October 1938.[123] In December 1938 the Reichsbahn prepared a revised five-year production plan for 6,000 locomotives, 10,000 carriages, 112,000 goods wagons and 17,300 lorries and trailers.[124] But, because the railway remained near the bottom of the regime's list of priorities for investment, the plan was never implemented.[125]

The outbreak of war surprised the Reichsbahn. As late as December 1938 the DR's directors expected no war.[116] Once it came, the board continued to hope that the war would be over soon, and admitted that the pace of events had outrun it since 1938.[127] The Reichsbahn's modernisation plans were shelved and its fast railcars consigned to the holding sidings as part of the general policy of allocating diesel fuel first to the Wehrmacht.[128] In October 1939 the DR's steel allocation was slashed by one-quarter, a clear indication of its low priority at the highest levels of the government.[129] Despite Göring's imprimatur, lack of raw materials, factory space and diesel fuel, in short the outbreak of war, brought the Reichsbahn's modernisation programme to an end.

IV

It is clear that the foundations for the modernisation of the Reichsbahn were laid under the Weimar Republic. From 1920 right up to 1939 the Reichsbahn's most important reasons for modernising were fiscal stringency and modal competition. The pace and direction of modernisation were not affected by the advent of the Nazi regime. The railway did not expand to meet the increased demands of rearmament and war. It was not subsidised by the government because it did not fit into the Nazis' vision of modernisation. Consequently, the Reichsbahn was excluded from the Four-Year Plan. It waged its struggle against road competition in virtual isolation. When the DR did contemplate war, it saw it as coming far in the future and resembling the First World War.

The dominant theme of this article is continuity. The problems confronting the Reichsbahn, and its responses to them, are part of a much broader set of transport trends discernible throughout the Western industrialised world since 1919. They appeared in Germany before the rise of the Hitler regime, and persist there to the present. The DR's struggle to modernise foreshadows the post-war history of the German Federal Railway (Deutsche Bundesbahn). The increase in road traffic, the competition from the airliner and the decline of heavy industrial traffic have not only continued but have accelerated. The railway's responses of introducing high-speed intercity passenger and goods trains, and cutting back personnel have been just as consistent.

Nazi propaganda presented a false image. National Socialism was not responsible for the triumphs of the Reichsbahn, and the DR was not mustering its considerable strength for war against the French or the Bolsheviks. Instead, the Reichsbahn was doggedly pressing on with its campaign to defeat the Lieblingskind of the Nazis, the automobile.

Notes

1 Geoffrey Freeman Allen, *Railways of the Twentieth Century*, (New York, 1983), p. 98.

2 Niederschrift, 4. Beirat, Deutsche Reichsbahn, 6 October 1937, p. 15, Bundesarchiv Koblenz, Registratur 2, file 31652. (Hereafter cited as BA R with the appropriate section and file numbers.)

3 Dorpmüller to Hitler, 30 December 1938, pp. 2–3, BA R43 II/183a, ff. 90–1.

4 International Military Tribunal, *Trials of the Major War Criminals at Nürnberg* (Nuremberg, 1946–48), XXXIII, Doc. 3787-PS, p. 157. See also Georg Thomas, *Geschichte der deutschen Wehr- und Rüstungswirtschaft 1919–1944/45* (Boppard, 1966), p. 147.

5 See, for example, Eugen Kreidler, *Die Eisenbahnen im Machtbereich der Achsenmächte während des Zweiten Weltkrieges* (Göttingen, 1975), pp. 31, 36–7, 46–7; Otto Wehde-Textor, 'Die Leistungen der Deutschen Reichsbahn im Zweiten Weltkrieg', *Archiv für Eisenbahnwesen*, LXXI, Jg (1961), pp. 5, 10; Hans-Erich Volkmann, *Das Deutsche Reich und der Zweite Weltkrieg* I (Stuttgart, 1979), p. 366; Klaus A. Schüler, *Logistik im Russland Feldzug: Die Rolle der Eisenbahn bei Plannung, Vorbereitung und Durchführung des deutschen Angriffs auf die Sowjetunion bis zur Krise vor Moskau im Winter 1941/42* (Frankfurt, 1987), pp. 59–88; Heinz Wehner, 'Der Einsatz der Eisenbahnen für die verbrecherischen Ziele des faschistischen deutschen Imperialismus im 2. Weltkrieg', Dresden, dissertation, Hochschule für Verkehrswesen, 1961, pp. 150–1.

6 For the purposes of this discussion, modernisation is defined as the introduction into regular use of new technologies and methods of organisation. The simple replacement of vehicles or facilities is not considered modernisation since obsolete technologies may be used to restore the serviceability of these items while allowing them to fall behind the state of the art.

7 The Deutsche Reichsbahn (DR) was founded on 1 April 1920. It was part of the Reich Transport Ministry and its finances were included in the Reich budget. On 12 February 1924 the Reichsbahn was separated from the national budget. Effective 1 October 1924 the Deutsche Reichsbahn-Gesellschaft was created to operate Germany's railways on behalf of the government, which retained ownership of the railways' real assets. From this point the abbreviation DRG was used. On 1 February 1937 the Reichsbahn was returned to the immediate control of the Reich government and DR became the official abbreviation. The use of abbreviations of the name of the Reichsbahn in the text reflects these changes.

8 The Reichsbahn lost 5,000 locomotives, 20,000 carriages and 150,000 goods wagons as a result of the stipulations of the Versailles Treaty. Kreidler, *Eisenbahnen im Machtbereich der Achsenmächte*, p. 15.

9 Niederschrift, 6. Sitzung, Verwaltungsrat, Deutsche Reichsbahn-Gesellschaft, 20 May 1925, p. 3, BA R2/31646.

10 Statistiches Reichsamt, *Statistisches Jahrbuch für*

das Deutsche Reich (Berlin, 1922), p. 106; Statistiches Reichsamt, *Statistiches Jahrbuch für das Deutsche Reich* (Berlin, 1932), p. 145.

11 Niederschrift, 33. Verwaltungsrat, Deutsche Reichsbahn-Gesellschaft, 26 November 1929, p. 8, BA R2/31647.

12 Aufzeichnungen, Arbeitsausschuss, Verwaltungsrat, Deutsche Reichsbahn-Gesellschaft, 21 March 1932, pp. 3–4, BA R2/23090a.

13 Manfred Zachcial, 'Das Ende eines Monopols, Eisenbahn und Kraftverkehr', *Zug der Zeit, Zeit der Züge*, Band 2 (Berlin, 1985), p. 663.

14 Niederschrift, 45. Verwaltungsrat, Deutsche Reichsbahn-Gesellschaft, 21, 22 September 1931, p. 11, BA R2/23090b.

15 Niederschrift, 46. Verwaltungsrat, Deutsche Reichsbahn-Gesellschaft, 24 November 1931, p. 13, BA R2/23090b.

16 Niederschrift, 33. Verwaltungsrat, Deutsche Reichsbahn-Gesellschaft, 26 November 1929, p. 13, BA R2/31647.

17 Hauptverwaltung der Deutschen Reichsbahn, *Hundert Jahre Deutsche Eisenbahnen* (Berlin, 1935), pp. 422, 431; Aufzeichnungen, Arbeitsausschuss, Verwaltungsrat, Deutsche Reichsbahn-Gesellschaft, 17 September 1928, p. 12, BA R2/31647.

18 DRG, HV, 'Das Wirtschaftsergebnis des Jahres 1927 nach den Ergebnissen der Betriebskostenrechnung mit Berücksichtigung der Kapital- und Schuldendienst', 22 Nr.Be.2, Berlin, 20 July 1928, pp. 15, 16, 21, BA R5/2455.

19 Deutsche Reichsbahn-Gesellschaft, *Dienstvorschrift für die Bearbeitung der Betriebsleistungsermittlungen in den Lochkartenstellen (VBB)* (Cologne, 1935), p. 35, BA RD98/25; Touve, Koch, 'Aufrechnung und Kontrolle durch Lochkartenmaschinen', *Die Reichsbahn*, III (1922), pp. 30–2.

20 The loan arranged by the Deutsche Wagenbau-Vereinigung was for 100 million RM in 1929. Among the institutions backing the loan were the Dresdener Bank, the Darmstädter-National Bank, and a number of Dutch and British banks. See the extensive correspondence on this matter in BA R5/2501. In 1930 and 1931, the Reichsbahn received 50 million RM from a consortium consisting of Siemens, BBC, AEG and Bergmann to electrify the Augsburg–Stuttgart line. The scheme was justified as a means of creating jobs during the depression. See the files on this in BA R5/2513.

21 Niederschrift, 17. Verwaltungsrat, Deutsche Reichsbahn-Gesellschaft, 29–30 March 1927, p. 8, BA R2/31646.

22 Parker Gilbert to Reichsminister der Finanzen, Berlin, 21 May 1928, BA R5/2599.

23 Reichsminister der Finanzen, Köhler to Dorpmüller, IV A 4112, Berlin, 23 May 1928, BA R5/2599; Gaston Leverve, Commissariat des Chemins de Fer Allemands, to Dorpmüller, Berlin, 28 May 1928, BA R5/2599.

24 Draft of letter from Dorpmüller to Gilbert 14 July 1928, BA R5/2599. The content of the letter had been co-ordinated with Gilbert in advance.

25 Deutsche Reichsbahn-Gesellschaft, *Statistische Angaben über die Deutsche Reischsbahn im Geschäftsjahr 1934* (Berlin, 1935), p. 87.

26 Walther G. Hoffmann, *Das Wachstum der deutschen Wirtschaft seit der Mitte des 19. Jahrhunderts* (Berlin, 1965), p. 143, Table 61.

27 The eight states that operated railways were Baden, Bavaria, Hesse, Mecklenburg-Schwerin, Oldenburg, Prussia, Saxony and Württemberg. The Prussian and Hessian systems were combined.

28 Aufzeichnungen, Arbeitsausschuss, Verwaltungsrat, Deutsche Reichsbahn-Gesellschaft, 22 September 1930, p. 8, BA R2/23090a; Niederschrift, 28. Verwaltungsrat, Deutsche Reichsbahn-Gesellschaft, 22 January 1929, p. 12, BA R2/31647.

29 Deutsche Reichsbahn-Gesellschaft, Hauptverwaltung, 'Ersparungsmassnahmen bei der Reichsbahn', 41 Kfl 13, Berlin, 1 May 1930, p. 4, BA R5/2539.

30 Hans-Wolfgang Scharf, *Eisenbahnen zwischen Oder und Weichsel. Die Reichsbahn im Osten bis 1945* (Freiburg, 1981), p. 189; DR, *Hundert Jahre*, pp. 263–4.

31 Niederschrift, 16. Verwaltungsrat, Deutsche Reichsbahn-Gesellschaft, 26 January 1927, p. 14, BA R2/31646; Aufzeichnungen, Arbeitsausschuss, Verwaltungsrat, Deutsche Reichsbahn-Gesellschaft, 24 November 1927, p. 7, BA R2/31646.

32 Max Leibbrand, 'Fortschritte und Probleme in der Rationalisierung des Reichsbahnbetriebes', Berlin 17–22 March 1930, p. 1, BA R5/2539.

33 DR, *Hundert Jahre*, p. 478.

34 Telegrammbrief, Deutsche Reichsbahn-Gesellschaft, Reichsbahn-Zentralamt, Wagendienst- und Verkehrsabteilung, an Reichsbahndirektionen, Wagenbüros, Oberbetriebsleitungen, 3301, Berlin, 25 May 1927, BA R5/2102; DR, *Hundert Jahre*, pp. 338–9, 349, 351, 373; Elfriede Rehbein, *et al.*, *Deutsche Eisenbahnen 1935–1985* (Berlin, 1985), p. 121; Deutsche Reichsbahn-Gesellschaft, *Geschäftsbericht der Deutschen Reichsbahn über das 1. Geschäftsjahr 1925* (Berlin, 1926), p. 69.

35 DRG, 'Ersparungsmassnahmen', p. 4.

36 Deutsche Reichsbahn-Gesellschaft, Hauptverwaltung, 'Bestand, Durchschnittsalter, Ausmusterung und Beschaffung von Dampflokomotiven, Personen-, Gepäck- und Güterwagen der Deutschen Reichsbahn während der letzten Jahren', 30.73a Fuv100, Berlin, 2 November 1935, p. 4, BA R5/2541.

37 Rehbein, *Deutsche Eisenbahnen*, pp. 126–277, 129–31.

38 *Ibid.*, p. 129.

39 DR, *Hundert Jahre*, pp. 160–1.

40 *Ibid.*, pp. 131–2.

41 DRG, 'Ersparungsmassnahmen', p. 2.

42 Leibbrand, 'Fortschritte', p. 3.

43 DRG, 'Ersparungsmassnahmen', p. 4.

44 DRG, 'Bestand', Anlage 4.

45 DRG, Hv, 'Stellungnahme der Hauptverwaltung der Deutsche Reichsbahn zu verschiedenen Zahlenvergleichen, zur Wirtschaftsrechnung und zu den Begründung in der Eingabe der Lokomotivindustrie vom 23. Januar 1932', 23 March 1932, pp. 4–5, BA R2/23090b; Homberger, *The Dawes Way*, 1/2 (January/February 1927), p. 5, BA R5/2046.
46 DRG, Hv, 'Stellungnahme der Hauptverwaltung der Deutsche Reichsbahn zu verschiedenen Zahlenvergleichen, zur Wirtschaftsrechnung und zu den Begründung in der Eingabe der Lokomotivindustrie vom 23. Januar 1932', 23 March 1932, Anlage 7, BA R2/23090b. From 1920 to 1929, the Reichsbahn ordered an average of 624·1 locomotives per year and spent 611 million RM on them.
47 DR, *Hundert Jahre*, p. 232; Rehbein, *Deutsche Eisenbahnen*, pp. 134–6; Scharf, *Eisenbahnen zwischen Oder und Weichsel*, p. 278.
48 DR, *Hundert Jahre*, pp. 135–6.
49 *Ibid.*, p. 135; Scharf, *Eisenbahnen zwischen Oder und Weichsel*, p. 278.
50 DR, *Hundert Jahre*, pp. 244, 249, 368.
51 Rehbein, *Deutsche Eisenbahnen*, p. 138.
52 Heinz R. Kurz, *Fliegende Züge. Vom 'Fliegenden Hamburgen' zum 'Fliegenden Kölner'* (Freiburg, 1986), pp. 5–8. See also Alfred B. Gottwaldt, *Schienenzeppelin* (Augsburg, 1972).
53 Rehbein, *Deutsche Eisenbahnen*, pp. 138–9.
54 Kurz, *Fliegende Züge*, pp. 12–25; Allen, *Railways of the Twentieth Century*, p. 97.
55 Allen, *Railways of the Twentieth Century*, pp. 87, 98; Heinz Wehner, in Rehbein, *Deutsche Eisenbahnen*, p. 152; DR, *Hundert Jahre*, p. 253; Kurz, *Fliegende Züge*, p. 21. The Fliegender Hamburger covered the 286·6 km (178·1 miles) in just 2 hours and 18 minutes.
56 Rehbein, *Deutsche Eisenbahnen*, p. 143; DR, *Hundert Jahre*, pp. 177–9.
57 Wehner, *Deutsche Eisenbahnen*, pp. 155–6.
58 DRG, 'Bestand', p. 8.
59 See the extensive correspondence on this matter in BA R2/2449.
60 Leibbrand, 'Fortschritte', p. 5.
61 DR, *Hundert Jahre*, p. 222; Kurz, *Fliegende Züge*, pp. 12, 14.
62 DR, *Hundert Jahre*, pp. 96, 100; Rehbein, *Deutsche Eisenbahnen*, p. 121–2.
63 Ursula-Maria Ruser, *Die Reichsbahn als Reparationsobjekt* (Freiburg, 1981), p. 356, Anlage 13. This is a reproduction of a document held in BA R43 II/187b. See also DRG, 'Bestand', Anlagen 7 and 10. The figure for expenditure for all vehicles in 1925 differs substantially in the latter document from that given in the Reichskanzlei document. Using it, the annual average is 138 million RM. See also Rehbein, *Deutsche Eisenbahnen*, p. 140; DR, *Hundert Jahre*, pp. 202, 205–6.
64 Leibbrand, 'Fortschritte', p. 6; DR, *Hundert Jahre*, p. 163.
65 Reichsbahn-Zentralamt für Bau- u. Betriebstechnik Berlin, 'Aufschreibung über ferngesteuerte Gleisbremsen...', December 1936, BA R5/2405; Leibbrand, 'Fortschritte', p. 5.
66 DRG, 'Ersparungsmassnahmen', p. 5; DR, *Hundert Jahre*, pp. 358, 360–1; Rehbein, *Deutsche Eisenbahnen*, p. 121.
67 Rehbein, *Deutsche Eisenbahnen*, p. 121.
68 DR, *Hundert Jahre*, p. 84.
69 Leibbrand, 'Fortschritte', pp. 4–5; DR, *Hundert Jahre*, pp. 318, 345.
70 DR, *Hundert Jahre*, p. 345.
71 Deutsche Reichsbahn-Gesellschaft, *Die Deutsche Reichsbahn im Geschäftsjahr 1925* (Berlin, 1926), p. 68.
72 Anger, Abt. III, 'Reichsbahn-Chronik 1934', 37 Aag 14, Berlin, 31 January 1935, BA R5/3337; DR, *Hundert Jahre*, p. 343.
73 Hans-Wolfgang Scharf, Friedhelm Ernst, *Vom Fernschnellzug zum Intercity* (Freiburg, 1983), pp. 749–52, Übersicht 5b. This was also substantially faster than the 1914 average of 58 km per hour. In 1932 it was 72·0 km per hour.
74 Leibbrand, 'Fortschritte', p. 3.
75 Compiled from Geschäftsberichte of the Reichsbahn for the years 1926–29.
76 Calculated from statistics in Deutsche Reichsbahn-Gesellschaft, *Die Deutsche Reichsbahn im Geschäftsjahr 1927* (Berlin, 1928), p. 45; Deutsche Reichsbahn-Gesellschaft, *Die Deutsche Reichsbahn im Geschäftsjahr 1929* (Berlin, 1930), p. 45.
77 Calculated from statistics in Deutsche Reichsbahn-Gesellschaft, *Die Deutsche Reichsbahn im Geschäftsjahr 1927*, pp. 44, 45.
78 Niederschrift, Ausserordentlicher Verwaltungsrat, Deutsche Reichsbahn-Gesellschaft, 29, 30 May 1934, p. 2, BA R2/23092.
79 Schüler, *Logistik*, p. 52 citing Der Staatssekretär in der Reichskanzlei, Aufzeichnung über die Unterhaltung zwischen dem Reichskanzler und dem General Direktor der Deutschen Reichsbahn-Gesellschaft, RK. 4285, 19 April 1933, BA R43 I/1053.
80 Wehner, *Deutsche Eisenbahnen*, p. 148; Niederschrift, 57. Verwaltungsrat, Deutsche Reichsbahn-Gesellschaft, 19 September 1933, p. 12, BA R2/23091.
81 Anton Joachimsthaler, *Die Breitspurbahn*, 3rd edn (Munich, 1985), p. 72.
82 Deutsche Reichsbahn-Gesellschaft, 'Umstellung des Reisezugdienstes auf Triebwagen', Drucksache Nr 795, 29 May 1934, p. 4.
83 *Ibid.*, p. 15.
84 *Ibid.*, p. 39.
85 DR, *Hundert Jahre*, p. 254; Scharf, Ernst, *Vom Fernschnellzug zum Intercity*, pp. 556–67.
86 Wehner, *Deutsche Eisenbahnen*, p. 152; Wolfgang Stoffels, Eberhard Krummheuer, *150 Jahre Deutsche Eisenbahnen* (Munich, 1985), p. 44. For a map of the system as of 1939 see Kurz, *Fliegende Züge*, p. 111. Chapters 3–5, 7 of Kurz provide detailed technical descriptions of the classes of high-speed diesel railcars built for the DRG after the Fliegender Hamburger.
87 Friedrich Witte, 'Zehn Jahre Reichsbahn-

Zentralamt Berlin und die Kriegslokomotiven 1935–1945', *Lok-Magazin*, XL (1970), p. 7.

88 Wehner, *Deutsche Eisenbahnen*, p. 152.

89 Anger, 'Reichsbahn-Chronik 1934'; Wehner, *Deutsche Eisenbahnen*, p. 152.

90 Rehbein, *Deutsche Eisenbahnen*, p. 136. The locomotives were the E 19 and the E 94.

91 Deutsche Reichsbahn, *Wirtschaftsplan für das Geschäftsjahr 1938* (Berlin, 1937), p. 25, Table 28; Deutsche Reichsbahn, *Wirtschaftsplan für das Geschäftsjahr 1943* (Berlin, 1943), p. 25, Table 28.

92 Wehner, *Deutsche Eisenbahnen*, p. 156.

93 Compiled from *Statistiche Angaben über die Deutsche Reichsbahn*, for the business years 1933 to 1938. See also Wehner, 'Einsätze', p. 38.

94 Scharf, Ernst, *Vom Fernschnellzug zum Intercity*, pp. 749–52, Übersicht 5b.

95 *Ibid.*, p. 666. These were the famous Schürzenwagen.

96 Witte, 'Zehn Jahre Reichsbahn-Zentralamt', p. 4. War locomotives were simplified, low-cost engines, intended to be built quickly in large numbers and to last only for the duration of the war.

97 Aufzeichnung, 55. Verwaltungsrat, Deutsche Reichsbahn-Gesellschaft, 3 May 1933, pp. 7–8, BA R2/23091. See also the files in BA R5/3432.

98 Wehner, *Deutsche Eisenbahnen*, p. 152; DR, *Hundert Jahre*, p. 89.

99 Scharf, *Eisenbahnen zwischen Oder und Weichsel*, pp. 217–18.

100 Niederschrift, 3. Beirat, Deutsch Reichsbahn, 6 July 1937, p. 19, BA R2/31652; Niederschrift, 5. Beirat, Deutsche Reichsbahn, 1 December 1937, p. 6, BA R2/31652; Niederschrift, 6. Beirat, Deutsche Reichsbahn, 26 January 1938, pp. 13–14, BA R2/31652.

101 Draft letter from Eltz-Rübenach to Lammers, September 1935, BA R5/2541. The letter was sent on 4 October 1935.

102 DRG, Hv to Eltz, 41 Kmbp (1935) 121, Berlin, 21 October 1935, BA R5/2541.

103 Schwerin von Krosigk to Eltz-Rübenach, Ve 1410/35-141c, Berlin, 20 September 1935, BA R5/2541.

104 Niederschrift, 1. Beirat, Deutsche Reichsbahn, 29 March 1937, pp. 8–9, BA R2/31652; Niederschrift, 4. Beirat, Deutsche Reichsbahn, 6 October 1937, p. 9, BA R2/31652; Bergmann, Vermerk, 'Besprechung mit General von Hanneken am 11. November 1938 über Erhöhung des Stahlkontingents der Reichsbahn', Berlin, 18 November 1938, BA R5/2123.

105 Hoffmann, *Wachstum*, p. 143, Table 61.

106 Deutsche Reichsbahn, *Geschäftsbericht der Deutschen Reichsbahn über das Geschäftsjahr 1937* (Berlin, 1938), pp. 55–56, Table XII; Deutsche Reichsbahn, *Geschäftsbericht der Deutschen Reichsbahn über das Geschäftsjahr 1938* (Berlin, 1939), pp. 60–1, Table XII.

107 Niederschrift, 33. Verwaltungsrat, Deutsche Reichsbahn-Gesellschaft, 26 November 1929, p. 13,
BA R2/31647; Niederschrift, 64. Verwaltungsrat, Deutsche Reichsbahn-Gesellschaft, 28 November 1934, p. 5, BA R2/23092; Wehner, *Deutsche Eisenbahn*, p. 159; Letter, Abteilung I to Dorpmüller, Berlin, 17 October 1935, BA R5/2541. Henning claims that the Reichsbahn enjoyed a net investment of 2 billion RM between 1933 and 1939. Friedrich-Wilhelm Henning, *Das Industrialisierte Deutschland 1914 bis 1978*, 5th edn (Paderborn, 1979), p. 162.

108 Derived from statistics found in the *Geschäftsberichte* and *Statistische Übersichte* of the Reichsbahn. On the Nazi regime's motorisation policy see R. J. Overy, 'Cars, roads and economic recovery in Germany, 1932–1938', *Economic History Review*, 2nd ser., XXVIII (1975), pp. 466–83.

109 Niederschrift, 58. Verwaltungsrat, Deutsche Reichsbahn-Gesellschaft, 28 November 1933, p. 3, BA R2/23092.

110 Aufzeichnung, Arbeitsausschuss, Verwaltungsrat, Deutsche Reichsbahn-Gesellschaft, 14 May 1935, p. 5, BA R2/23092; Abt I to Dorpmüller, Berlin, 17 October 1935, BA R5/2541.

111 Niederschrift, 64. Verwaltungsrat, Deutsche Reichsbahn-Gesellschaft, 28 November 1934, p. 8, BA R2/23092.

112 Zachcial, *Zug der Zeit, Zeit der Züge*, Band 2, p. 663.

113 Niederschrift, 8. Beirat, Deutsche Reichsbahn, 2 July 1938, pp. 17–18, BA R2/31652.

114 Henning, *Das Industrialisierte Deutschland*, p. 112.

115 USSBS, *The Effects of Strategic Bombing on German Transportation* (Washington, 1947), p. 6. Overy cites a private estimate that by 1935 60 per cent of 'private journeys' were made by road. This seems high. Overy, 'Cars, roads and economic recovery', p. 481.

116 Wehner, *Deutsche Eisenbahnen*, p. 159; Deutsche Reichsbahn-Gesellschaft, *Geschäftsbericht der Deutschen Reichsbahn-Gesellschaft über das 12. Geschäftsjahr 1936* (Berlin, 1937), pp. 7, 17.

117 Niederschrift, 12. Beirat, Deutsche Reichsbahn, 9 May 1939, pp. 6–7, BA R2/31652.

118 Niederschrift, 56. Verwaltungsrat, Deutsche Reichsbahn-Gesellschaft, 4 July 1933, p. 9; Niederschrift, 11 Beirat, Deutsche Reichsbahn, 8 March 1939, p. 10, BA R2/31652.

119 RVM, E-Abt I to Abt II, 'Fahrzeugprogramm für mehrere Jahre', 10 Vwah 29, Berlin, 13 December 1937, BA R5/2090.

120 Bergmann, Abt III to Abt I, II, IV, Gruppe L, 'Bedarfsprogramm für neue Fahrzeuge für die 4 Jahre 1939 bis 1942, getrennt nach Fahrzeuggruppen, Stückzahlen und Gesamtkosten, Fall A', 30 Fef 58, Berlin, 31 December 1937, BA R5/2123.

121 RVM, E-Abt III, Kühne, to Abt II, 'Sofortprogramm 1939', 30 Fef 60, Berlin, 19 January 1938, BA R5/2090.

122 RVM, Bergmann, Abt III to Abt II, 'Beschaffungsplan für mehrere Jahre', 30 Fef 66, Berlin, 29 July 1938, BA R5/2090. See also Ref 23, Baüg 42,

'Übersicht I. Bedarfsprogram für neue Fahrzeuge für die 4 Jahre 1940 bis 1943 getrennt nach Fahrzeuggattungen und Stückzahlen', BA R5/2090.

123 'Conference at the General Field Marshall Goering's at 1000, 14 October 1938, in the Reich Air Ministry', Top Secret, Doc. 1301-PS, *Nazi Conspiracy and Aggression*, III, p. 901; 'Material for the Conference with Goering on 25 November 1938 (General Keitel, Brig. Gen. Thomas)', W.Wi Id, 27 October 1938, *Nazi Conspiracy and Aggression*, III (Washington, DC, 1946–48), pp. 904–6; Niederschrift, 10. Beirat Deutsche Reichsbahn, 7 December 1938, pp. 10, 12, 15–16, BA R2/31652.

124 Niederschrift, 10. Beirat, Deutsche Reichsbahn, 7 December 1938, pp. 10, 12, 15–16.

125 Karl-Heinz Ludwig, *Technik und Ingenieure im Dritten Reich* (Düsseldorf, 1979), p. 186.

126 Niederschrift, 10. Beirat, Deutsche Reichsbahn, 7 December 1938, p. 23.

127 Niederschrift, 15. Beirat, Deutsche Reichsbahn, 30 January 1940, p. 34, BA R2/31653; Niederschrift, 16. Beirat, Deutsche Reichsbahn, 28 May 1940, p. 13, BA R2/31653.

128 Kurz, *Fliegende Züge*, pp. 25, 48, 82, 112, 193; Scharf, Ernst, *Vom Fernschnellzug zum Intercity*, p. 90.

129 DR, E-Abt, 'Festlegung desjenigen Teiles der noch unerledigten Aufträge auf Wagen aller Art, aus den Beschaffungsprogrammen bis einschliesslich 1939 Zusatz, der zu Ende geführt werden soll', Berlin, 17 October 1939, p. 1, BA R5/2123.

10 A. D. Chandler's 'visible hand' in transport history: a review article

G. CHANNON

The subject of Chandler's study[1] is an institution: the modern business enterprise which emerged to dominate American economic life through a process of challenge and response in the period 1850–1920. The challenge was found in the growth of the market and related technological advances which made first the processes of distribution, then production, more complex, straining traditional market mechanisms and creating the need for administrative control and the co-ordination of resources. The response, complete in its essentials by 1920, involved the displacement of the small, specialised family or household firm by the multi-unit business enterprise run by a new class – managers – who in turn shaped subsequent developments. In all but final demand, therefore, the 'invisible' market was replaced by the 'visible' hand of management which appraised, allocated and administered resources in the private sector. This outcome, but not necessarily the historical processes that led to it, would appear to be characteristic of advanced industrial capitalism throughout the world.[2]

The purpose of *The Visible Hand* is to explain how this transformation came about in the United States. In Chandler's explanation transport and communications, especially railroads, are given a central role. It was in the railroads that the visible hand was first revealed, and it was the railroad system that was primarily responsible for the advent of large-scale distribution arrangements and the encouragement of high-volume industrial production.

I

Just over a quarter of the book's 500 pages of text are therefore devoted to the revolution in transport and communication. Challenged first by the requirements of safety, then by the volume, speed and complexity of traffic distributed over a widening area, railroad leaders were forced, from 1850, to evaluate their administrative structures and systems. There were a variety of organisational responses at this time, but Chandler attributes the best practice to the Pennsylvania, and roads that followed it, in the adoption of the decentralised, divisional structure and new practices in financial, cost and capital accounting. These innovations were mainly the work of salaried managers like J. Edgar Thomson (Pennsylvania), D. C. McCallum (Erie) and B. Latrobe (Baltimore & Ohio) — all trained civil engineers who, with the growing complexity and volume of managerial decisions, quickly captured much of the power that was legally vested in the shareholders and directors.

Chandler argues that while these administrative changes improved the performance and productivity of individual roads, before the Civil War the challenge and opportunity for through traffic in a truly national network was not being met. As a result, transhipment costs were high both to the roads and to the traders. The type of co-operation required — the standardisation of equipment and operating procedures — was 'an entirely new phenomenon' (p. 123) in inter-firm relations. In one sense it proved highly successful, for by the 1880s a rail consignment could be moved from one part of the country to another without a single transhipment. However, as the network expanded and became physically and organisationally more integrated, competition increased for traffic carried on parallel routes. To control such competition managers first sought informal alliances, followed by formal federations, such as the Southern Railway and Steamship Association, a large-scale pool which was organised in 1875 by Albert Fink, the leading figure in the cartelisation movement. By the early 1880s, though, administrative problems, disagreements over the pooling and allocating of income and above all 'the relentless pressure of high constant costs' (p. 143) resulted in the undermining of cartels. Often it was the middle managers in traffic departments, whose success depended on obtaining customers, who secretly cut rates and when the traffic fell off were also the ones who recommended to top management that the agreed rate structure should be abandoned. The final solution, or so it appeared, was to build self-sustaining 'megacorps'. The basic motive was defensive; the outcome was over-capacity and, in the 1890s, more bankruptcy in the industry than before or since.

Chandler traces in detail how the interplay between top managers, speculators, eastern capitalists and investment bankers shaped the strategies and destinies of railroads in the era of system-building. Generally it was the salaried managers who were able to convince financiers of the need to build the systems, who managed the strategy of growth and who, once competitive pressure was largely removed (by the early years of the twentieth century), settled into an increasingly bureaucratic regime of railroad administration. Financiers provided the immense amounts of capital that growth required, played an active role in consolidations, and in their administrative reorganisations adopted the centralised, departmental structure (familiar to students of British railway history) in preference to the decentralised one. By the early twentieth century nearly all railroads were using this type of structure. They were the survivors in an industry controlled by a small number of 'managerial' enterprises.[3] 'The American railroad enterprise,' writes Chandler, 'might more properly be considered a variation of managerial capitalism than an unalloyed expression of financial capitalism' (p. 187).

By 1900 the railroads, together with steamship lines, urban traction, the postal service and the telegraph (increasingly the telephone), had ceased to be the creators and shapers of new economic institutions in the United States (for example, the capital market, the managerial enterprise, the regulatory commission). They had played out the historic role of helping to create a national market which permitted the spread of the visible hand to other areas of the economy, and had ceased to be the centres of political and ideological controversy. These roles were now assumed by the large, integrated, multi-unit firms that came to dominate American industry. The remaining chapters of *The Visible Hand*, which need not detain us here, are devoted to an explanation of the process by which such firms emerged and how they were administered. Producing and selling goods, of course, differed from transporting them, so the new industrial giants had to absorb, modify and rethink the methods of finance, administration, inter-firm co-operation and regulation in which the railroads, the nation's first big business, had been pioneers.[4]

II

In choosing institutional change as his theme Chandler is part of a revival – a 'new institutionalism' – of a concern that dominated economic history when it came to life in Germany and England a century ago.[5] While historians have traditionally shown an interest in the institutions within and between which human action occurs, be the institution social, political or military, economic historians have in recent decades tended to take the institutional 'framework' as given. Now other social scientists, notably economists such as J. K. Galbraith, P. A. Baran and P. M. Sweezy, R. L. Marris, E. T. Penrose and W. J. Baumol have studied more closely the functioning and policies of the modern business enterprise, leading them to question both the traditional notion of the firm as a unit of production and the theory of the firm as a theory of production.[6] Chandler, however, is the first scholar to trace the story of the rise of the modern business enterprise and the managerial capitalism that accompanied it.

His method, confidently stated at the outset, is that of the traditional historian: 'The data have not been selected to test and validate hypotheses or general theories' but it is the record of the past 'that provides the basis for the generalisations presented' (p. 6). Chandler therefore turns his back on the 'new' economic history and where he is forced to consider it, in the matter of the railroads and the growth of the market, responds in the manner of the traditional historian by drawing attention to the unrealistic assumptions and unrepresentative data used by its leading practitioner, R. W. Fogel.[7]

While *The Visible Hand* is essentially a work of synthesis, to a large extent it grows out of Chandler's own research and insights and the research of others that he has inspired. Clearly a central place must be given to his *Strategy and Structure: Chapters in the History of Industrial Enterprise* (Cambridge, Mass., 1962), in which he examined the large industrial enterprise as an institution.[8] He focused on the spread of the 'decentralized structure' after 1920 and showed in great detail how enterprises, often painfully, made administrative responses to the challenges posed by a change of corporate strategy. In that work, as in *The Visible Hand*, he argued that railroads stimulated the development of the national market and thus the growth of industrial concerns.

Chandler had earlier studied patterns of railroad finance before the Civil War and written a biography of Henry Varnum Poor, editor of the *American Railroad Journal*, railway analyst and reformer, who was a prominent figure in the mid-nineteenth-century railroad world.[9] In 1965 he published *The Railroads: the Nation's first Big Business* (New York, 1965). This contained documentary extracts introduced by the author in short, stimulating essays, organised round the railroad themes which are to be found in *The Visible Hand*. However, in *The Visible Hand* Chandler has additionally brought to bear on those themes the recent and not so recent monographic work of other railroad historians. Thus he has incorporated and generalised from the conclusions of J. A. Ward, who found that after 1852 career managers dominated the strategic decision-making of the Pennsylvania.[10] He also puts to good use the findings of T. C. Cochran and S. Morris which relate to the origins, careers and roles of railroad leaders.[11] On the development of particular railroads and their interaction, Chandler draws together the research of numerous authors, notable among them J. Grodinsky, E. C. Kirkland, M. Klein, A. M. Johnson and B. E. Supple, P. W. MacAvoy, A. Martin and J. F. Stover.[12] Finally, on the subject of regulation, he considers and rejects G. Kolko's view that, having failed to bring stability to the industry through their own efforts, most railroad leaders supported the Act of 1887 to regulate inter-state commerce and that they indeed received from it the benefits of stability they had wanted.[13] Instead Chandler follows the reappraisal of A. Martin, who argues that as early as the mid-1880s leading railroad men and investment bankers looked forward to the rapid consolidation of the railroads into self-sustaining systems.[14] If they expected any assistance from the government it was in making pooling contracts enforceable by law. The Act of 1887, however, made pooling illegal, and Supreme Court interpretations in 1897 and 1898 of the Sherman Anti-trust Act (1890) meant that even unenforceable agreements to uphold official tariffs in the absence of pooling were illegal. When these possibilities were closed the only alternative was formal consolidation and 'communities of interest' between the remaining roads, which stabilised the industry after the depression of the early 1890s.

Chandler's synthesis of research on railroad regulation has been quoted at some length because it typifies an interpretation that runs all through *The Visible Hand*. While he is more alive to the possible impact on corporate strategy and structure of public policy than in 1962, he continues to minimise the role of this factor, stressing instead the diversity of business behaviour in the light of changing market and technical forces. In the end, although Chandler has his managerial villains as well as heroes, there can be little doubt that he believes the process he describes to have been a rational one which conferred net social benefits through, or as well as, net private benefits. He fleetingly admits, for example, that system-building in the 1880s may have led to waste but quickly adds, 'In time, however, most of the new roads became fully used' (p. 148). The historian who is interested in welfare will find little here: labour conditions, income distribution, environmental questions and many others fall outside Chandler's brief.

III

Chandler, then, is the intellectual heir to a tradition that has its roots in the studies of institutionalists such as Weber, Schmoller, Veblen, and Berle and Means. As such, in *The Visible Hand* he reveals the limitations, strengths and possibilities (not least to students of British transport history) of that tradition.

It is clear that a central part of his thesis of institutional change in the American business system depends upon proving the reality of a long causal chain. This links organisational change in the railroad industry with improved railroad efficiency, reduced railroad rates, and market expansion, which permitted the institutional innovations in distribution and production that occurred in the later part of the nineteenth century. An appealing thesis, it rests uneasily upon a set of quantitative assumptions and inferences, the validity of which Chandler chooses not to demonstrate in detail. For example, in discussing railroad efficiency he states, 'Some productivity increases surely came from the administrative arrangements...' (p. 133). But to what extent? Here Chandler quite rightly asserts the role of a missing element in the traditional theory of the firm: administrative change. Its significance, however, would seem to depend upon a more sustained theoretical and statistical analysis than he is able to offer. Such analysis would seek out the complex relationships between administrative structure, process (both well defined by Chandler) and performance, within the individual roads and in the industry as a whole.[15]

The final link in the causal chain, the stimulation of the market by the railroads, requires him to reject the conclusion of Fogel's study that the railroads were a dispensable element in the economic growth of nineteenth-century America.[16] He confines his rejection to a footnote (p. 531), where he rehearses some of the major criticisms offered by leading 'new' economic historians, such as P. David, S. Lebergott and P. D. McClelland.[17] He finds that Fogel's handling of inventory costs is particularly misleading. Fogel's hypothetical alternative to the railroads, an expanded waterway system, would, in Chandler's view, have resulted in much higher inventory costs than Fogel admits, which would have acted as a barrier to the expansion of factory production.[18] The disagreement with Fogel is not, however, explicitly aired in the text, where Chandler seeks to show through *a priori* reasoning and a mass of qualitative evidence that the railroads were indispensable. The flavour of his method is captured in the following:

> The close cooperation between the managers of the first modern multi-unit enterprises in the United States contributed impressively to increasing the speed and regularity of transportation and decreasing its costs. And . . . it was the economy and velocity of transportation that provided the basic underpinnings of the institutional changes in American production and distribution . . . [page 133]

Those historians who count will surely ask: maybe, but to what extent?

IV

None the less, Chandler has written a brilliant study of administrative change, drawing upon his unsurpassed knowledge of the internal histories of American firms. Already the strategy–structure thesis published in 1962 has directly influenced studies of corporate development and organisation both inside and outside the United States. It has been exposed, for example, to the critical while admiring pens of several leading historians of British business.[19] Railway historians in Britain, however, have remained outside the debate. This may reflect the relative isolation of transport history from business history in this country, compared with the United States, where the history of transport is often regarded as an integral part of the subject matter of business and entrepreneurial history.[20] Chandler is perhaps the leading exponent of this relationship. In conclusion it is therefore proposed to consider briefly and provisionally the relevance to the British experience before 1920 of his analysis of railroad development.

Chandler's typology for the development of American railroads, starting with an emphasis on internal administrative problems, then inter-road administration and rivalry and finally, when cartels had failed, on system-building, cannot readily be transposed to Britain. The elements were undoubtedly present but the sequence was much more compressed and influenced to a greater degree by political pressures. For in Britain railway leaders had to contend from the early years with the problems of internal and inter-company administration, competition and political interference. For example, the development, from 1842, of the Clearing House to promote inter-company exchanges of traffic[21] occurred in the context of an industry in which, through new construction and amalgamation, the level of concentration was growing rapidly. Already by 1850 the four leading companies were taking 41·7 per cent of gross traffic receipts in England, Wales and Scotland, a share that was only slightly exceeded (47·0 per cent) seventy years later — shortly before that other visible hand, the State, intervened to transform the structure of the industry through the Railways Act of 1921.[22] Oligopolistic competition, which Chandler notes as a characteristic of the mature American railroad system, was therefore an early feature in Britain, where it took various forms over time: price competition which reached a peak in the 1850s, followed by competition for urban sites and finally, after 1870, service competition.

In Britain corporate strategy was not simply a matter of market calculation (as Chandler tends to argue for the United States). Throughout their history British railways have had to anticipate and react to the inconsistent and seemingly arbitrary requirements of the State. From the early years the competitive struggle was often carried into Parliament, where it might cut across any sense of co-operation that existed among the 'railway interest'.[23] It was the threat of parliamentary disapproval, especially after the rejection of several major amalgamation bills in the early 1870s, that appears to have effectively removed the option of growth through large-scale amalgamations and was therefore a major influence on the relatively stable level of concentration until 1921. This left railway leaders to rely on less formal and often less stable methods of regulating competition between themselves, such as the pricing and pooling agreements which proliferated in the last quarter of the nineteenth century.[24] After 1900, however, the major companies, as some of their American counterparts had attempted in the early 1880s, sought official approval of wide-ranging pooling agreements, which were designed (and in a small way may have helped) to arrest the recent decline in their financial fortunes. But, as P. J. Cain has demonstrated, such was the intensity of opposition from Parliament, the traders and their own employees that the companies removed their agreements from the public gaze and tried to continue secretly.[25] Thus on the eve of the first world war British railways found themselves constrained by the State in their attempts to regulate oligopolistic competition: large-scale amalgamations were unlikely to get parliamentary approval, while pooling and similar devices remained as in the nineteenth century: underground and possibly, in certain aspects, illegal.[26]

It is apparent, therefore, that political factors played a more continuous and important conditioning role in Britain than in the United States in what has aptly been called 'high-order' entrepreneurial activity, that is, activity which altered the overall pattern of a company's activities.[27] Such factors, however, in both countries influenced 'low-order' entrepreneurial activity, that is, operating methods, including pricing, in ways that appear to have had a significant effect on the profitability and productivity of the respective railway systems at the turn of the nineteenth century. R. J. Irving, in an examination of the British system as a whole (in which he often generalises from his work on the North Eastern, a leader of advanced management), and T. R. Gourvish, in a study of companies that are usually cited as exemplars of alleged 'worst practice', have reached similar conclusions.[28] In the 1880s and 1890s poor profitability was the outcome of 'inappropriate' operating policies, conditioned as they were by a hostile political environment which reduced the options open to managers. Irving argues that, once government control over rates was finally resolved (in 1899), managements were able to promote operating efficiency, especially in their freight departments.[29] Even so, they still had to contend with official resistance to a wider-ranging set of 'high-order' proposals for the structure of the industry.

In the United States the surviving giants, having stabilised the railroad system through their own efforts, opposed bitterly (and unsuccessfully) proposals for increasing the power of the Interstate Commerce Commission to fix rates. According to A. Martin, in a study not cited by Chandler, the unwillingness of the ICC to grant rate increases during a period of policy uncertainty, starting with the Hepburn Act (1906), led to capital undernourishment.[30] The flow of investment funds was stemmed at a time when railroad traffic output was increasing, and thus the way was opened to a collapse in the profitability of operations after 1911.

The railways were Britain's first big business too. Like their American counterparts, the larger companies came to display the familiar organisational features of bureaucratic institutions, including elaborate hierarchies, a reliance on rules and recognised procedures and administrative structures devised according to certain 'rational' principles. No doubt the requirements of safety and the dispersed, complex nature of their activity presented British railway leaders with similar challenges, even if their responses may have differed in detail. Administrative structures, for example, were typically of the centralised, departmental type, under the co-ordinating control of the general manager, without a tier of middle general management and involving, therefore, the isolation of the revenue-earning from the spending departments. Similar centralised structures, according to Chandler, had by the end of the nineteenth century displaced the decentralised structure as the dominant model in America. There, as we have seen, the requirements of financiers were the main influence. In Britain modern historiography has not, at least for the period before 1920, sought to explain why the departmental system was dominant. Bonavia is the only author to take an overall view of the organisation of British railways.[31] While he concentrates on the period after nationalisation and does not seek to explain why the traditional system arose, he nonetheless offers clues as to why it survived and how it may have adversely affected business performance. It often led, for example, to a conflict between the goals and interests of departments and to a failure to isolate the true net revenue position of particular operations.[32] Once established, the system had a tenacious hold which was not threatened before the inter-war years.[33]

Chandler offers the British railway historian some useful pointers to the possible symbiotic relationships between the administrative structures of railroads and strategy, noting, for example, the complex interactions between top managers, middle managers and financiers, each of whom had an identifiable and different corporate goal. Historians of British railway companies have only recently begun to investigate these issues. T. R. Gourvish, for example, tries to establish both the facts and the significance of the administrative innovations introduced by the first general manager of the London & North Western, Mark Huish.[34] He gives an account of the management structure that Huish set up and describes the systems by which he sought to identify, collect and analyse information as a basis of rational decision. He also explains how ultimately Huish was unable to satisfy his masters, who were especially critical of his conduct of inter-company relations and who forced him to resign in 1858.[35] Huish's fate stands as a warning that management ascendancy, as in the United States, was not necessarily a straightforward linear process. Unlike Chandler, Gourvish attempts, through the construction of time series data for the London & North Western, and financial information from this and several other major companies, to assess management performance.

Undoubtedly studies of this kind present the historian with considerable empirical problems — for example, the discovery of where in the company 'high' and 'low'-order decisions were made; the identification of who controlled information and (possibly) policy options; and the construction, from information designed for other purposes, of acceptable measures of performance. Yet if the spread of the visible hand in British railways is to be described and understood such problems must be tackled.

Whether or not the railways, through their effects on the market and their business methods, promoted the spread of the visible hand in other sectors of the economy (as Chandler contends they did in United States) is still something of an open question. In the late nineteenth century there was a rapid increase in nation-wide sales of consumer goods, often branded and requiring new methods of marketing and distribution. There followed at the turn of the century the emergence of the first large-scale firms in industry — the products of a merger boom which was activated by defensive market considerations, rather than a desire for improved efficiency, and facilitated by changes in the methods of financing and promoting limited companies.[36] Often the new creations, which were much smaller than their American counterparts (or the leading British railways), were federations of family firms. Not until the inter-war years was there a significant move towards the managerial enterprise in industry. According to Chandler the final transformation of business as an institution, to date, is the emergence and domination of the multi-divisional firm. In Britain such firms developed rapidly from the 1950s (several decades behind the United States) to dominate the list of the top hundred industrial enterprises by 1970.[37]

Several reasons may be suggested for caution in attributing the origins of these changes in British enterprise to the market effects of the railways. Britain is a small country which at the eve of the railway age was already well served by integrated commodity markets, adequate waterborne transport and the highest levels of urban concentration in the world. Yet the large-scale firm did not emerge until fifty years later. Even allowing that the railways helped to promote the extension of the market,[38] providing a context for inter-firm competition and large-scale production and distribution, the response was often the informal cartel and/or product differentiation, each of which helped to preserve the small family and entrepreneurial firm.[39] If these comments have any validity, then a corollary might be that the business methods developed by the railways — for example, in depreciation accounting, procedures for appraisal and planning, and line and staff management — would have appeared inappropriate in industry before the turn of the century or indeed later. When large-scale firms began to confront the kinds of problems that had prompted these methods, the more enterprising of them could turn to the United States for solutions. However, the possibility of a transference of business methods through the participation of former railway executives in other sectors cannot yet be ruled out.[40]

V

Comparative history has been seen as a useful way of providing insights into the process of past economic growth and development; it may prove equally valuable in the study of modern business institutions, including railways, if used with care. In this review article a modest, preliminary survey has been undertaken of some of the possibilities that might interest the historian of British railways. Chandler in fact, at the end of *The Visible Hand,* points the way forward for research in this direction, defining the object and benefit of such an exercise as follows:

> Describing and analysing the history of the new institution [modern business enterprise] and the ways in which it has carried out its basic functions in different nations can help to define the organizational imperatives of modern economies and reveal much about the ways in which cultural attitudes, values, ideologies, political systems, and social structure affect these imperatives. [page 500]

As might be expected, he does not himself attempt to provide an overall theory of business development under modern capitalism.[41] To the last he is loyal to his chosen method: generalisation from an accumulation of particular instances, awareness of the exceptional and unrepeatable, and the avoidance of a deterministic interpretation. It is a method, however, that eventually yields a fascinating thesis about the development of the American business system, a thesis which places the railroad at the centre of the story. As such, *The Visible Hand* should be a source of inspiration to transport historians on both sides of the Atlantic.

Notes

1 A. D. Chandler junior, *The Visible Hand: the Managerial Revolution in American Business* (Cambridge, Mass., and London, 1977).

2 On Britain, for example, see L. Hannah, *The Rise of the Corporate Economy* (1976).

3 By 1893 thirty-three railroads with a capitalisation of $100 million or more operated 69 per cent of the railroad mileage. The largest systems in 1893 were almost the same as those in 1906 and 1917, although at the latter dates overall concentration was higher. *The Visible Hand,* tables 3 and 4, pp. 168–9.

4 Chandler pays particular attention to the decentralised forms of organisation pioneered by the railroads, which offered a model to the large new industrial corporations.

5 The revival is represented by L. E. Davis and D. C. North, *Institutional Change and American Economic Growth* (Cambridge, 1971), and D. C.

North and R. P. Thomas, *The Rise of the Western World* (Cambridge, 1973).

6 J. K. Galbraith, *The New Industrial State* (1967); P. A. Baran and P. M. Sweezy, *Monopoly Capitalism* (Harmondsworth, 1968); R. L. Marris, *The Economic Theory of 'Managerial' Capitalism* (1964); E. T. Penrose, *The Theory of the Growth of the Firm* (Oxford, 1959); W. J. Baumol, 'Entrepreneurship in economic theory', *American Economic Review*, LXXX (1967).

7 R. W. Fogel, *Railroads and American Economic Growth: an Econometric History* (Baltimore, 1964). For a further discussion see section III of this article.

8 Note also A. D. Chandler and S. Salsbury, *Pierre S. du Pont and the Making of the Modern Corporation* (New York, 1971).

9 'Patterns of railroad finance, 1830–1850', *Business History Review*, XXVIII (1954); *Henry Varnum Poor: Business Editor, Analyst and Reformer* (Cambridge, Mass., 1956). Also see Chandler's 'The railroads: pioneers of modern corporate management', *Business History Review*, XXXIX (1965).

10 J. A. Ward, 'Power and accountability on the Pennsylvania Railroad, 1846–1878', *Business History Review*, XLIX (1976).

11 T. C. Cochran, *Railroad Leaders, 1843–1899: the Business Mind in Action* (Cambridge, Mass., 1953); S. Morris, 'Stalled professionalism: the recruitment of railway officials in the United States, 1885–1940', *Business History Review*, XLVIII (1973).

12 J. Grodinsky, *The Iowa Pool* (Chicago, 1950); also *Jay Gould: his Business Career, 1867–1892* (Philadelphia, 1957), and *Transcontinental Strategy, 1869–1893* (Philadelphia, 1962); E. C. Kirkland, *Men, Cities and Transportation: a Study in New England History* (Cambridge, Mass., 1948); M. Klein, 'The strategy of southern railroads', *American Historical Review*, LXIII (1968); A. M. Johnson and B. E. Supple, *Boston Capitalists and Western Railroads* (Cambridge, Mass., 1967); P. W. MacAvoy, *The Economic Effects of Regulation: the Trunk Line Railroad Cartels and the Interstate Commerce Commission before 1900* (Cambridge, Mass., 1965; a very detailed study of pricing, using advanced economic techniques); A. Martin, *James J. Hill and the Opening of the Northwest* (New York, 1976); and J. F. Stover, *Railroads of the South* (New York, 1961).

13 G. Kolko, *Railroads and Regulation, 1877–1916* (Princeton, N. J., 1965).

14 A. Martin, 'The troubled subject of railroad regulation in the Gilded Age – a reappraisal', *Journal of American History*, LXI (1974).

15 The conceptual problems are explored by B. W. E. Alford in 'Entrepreneurship, business performance and industrial development', *Business History*, XIX (1977), and C. E. Harvey, 'Business history and the problem of entrepreneurship: the case of the Rio Tinto Company, 1873–1939', *Business History*, XI (1980).

16 Fogel, *passim*.

17 P. David, 'Transportation and economic growth: Professor Fogel on and off the rails', *Economic History Review*, 2nd series, XX (1969); S. Lebergott, 'United States transport and externalities', *Journal of Economic History*, XXVI (1966); and P. D. McClelland, 'Railroads, American economic growth and the new economic history: a critique', *Journal of Economic History*, XXVIII (1968).

18 Time lost in slow movement and the closing down of waterways in the winter months would have contributed to high inventory costs.

19 E.g. L. Hannah (ed.), *Management Strategy and Business Development: a Historical and Comparative Study* (1976).

20 Note, in this respect, the large number of articles on transport history that have appeared in *The Business History Review*, the journal published by the Graduate School of Business Administration at Harvard.

21 P. S. Bagwell, *The Railway Clearing House in the British Economy, 1842–1922* (1968).

22 Calculated from the *Railway Returns* and incorporated in this reviewer's forthcoming article 'The reorganisation of British railways under the 1921 Railways Act: the case of the Great Western', *Business History Review*, LV (1981).

23 On the railway interest see G. Alderman, *The Railway Interest* (Leicester, 1973).

24 See the examples cited in G. Channon, 'Pooling agreements between the railway companies involved in Anglo-Scottish traffic, 1851–69', unpublished University of London Ph.D. thesis (1975), especially pp. 427–37.

25 P. J. Cain, 'Railway combination and Government, 1900–1914', *Economic History Review*, 2nd series, XXV (1972).

26 Channon, 'Pooling agreements', pp. 338–51.

27 Harvey, *loc. cit.*

28 R. J. Irving, 'The efficiency and enterprise of British railways, 1870–1914: an alternative hypothesis', *Economic History Review*, 2nd series, XXXI (1978), and *The North Eastern Railway Company, 1870–1914* (Leicester, 1976); T. R. Gourvish, 'The performance of British railway management after 1860: the railways of Watkin and Forbes', *Business History*, XX (1978). Gourvish has a good summary of the debate.

29 Irving, *loc. cit.*, pp. 279–81.

30 A. Martin, *Enterprise Denied: Origins of the Decline of American Railroads, 1897–1917* (New York, 1971).

31 M. R. Bonavia, *The Organisation of British Railways* (Shepperton, 1971).

32 *Ibid.*, pp. 153–4.

33 Note the limited changes in the organisation of the North Eastern Railway in the early twentieth century, although there was some management enthusiasm for the system used by the Pennsylvania Railroad. Irving, *op. cit.*, pp. 256–7.

34 T. R. Gourvish, *Mark Huish and the London and North Western Railway: a study in Management* (Leicester, 1972).
35 *Ibid.*, p. 166.
36 P. L. Payne, 'The emergence of the large-scale company in Great Britain', *Economic History Review*, 2nd series, XX (1967), p. 521. For details of industrial concentration see L. Hannah, *The Rise of The Corporate Economy*, p. 216.
37 A. D. Chandler, 'The development of modern management structure in the U.S. and U.K.', in Hannah (ed.), *Management Strategy*, p. 23.
38 G. R. Hawke, in *Railways and Economic Growth in England and Wales, 1840–1870* (Oxford, 1970), tries to establish the contribution of railways to the extension of domestic markets through the study of four commodities: wheat, meat, livestock and minerals. For minerals and meat the decline in c.i.f. prices brought about by railways seems substantial. For critiques of Hawke's method and use of data see P. O'Brien, *The New Economic History of the Railways* (1977), and C. H. Lee, *The Quantitative Approach to Economic History* (1977), pp. 81, 83–5, 87.
39 Payne, *loc. cit.*, pp. 524–6. The character of the market, with its skewed income distribution, may have been a factor here. But note the ability of entrepreneurs to determine their market environment.
40 T. R. Gourvish, 'A British business elite: the chief executive managers of the railway industry, 1850–1922', *Business History Review*, XLVII (1973), pp. 311–14. The role of directors as a channel for diffusion in *both* directions is a further possibility.
41 Note his excursion into comparative history (US and UK) in Hannah (ed.), *Management Strategy*, pp. 23–51, and 'The growth of the transnational industrial firm in the United States and the United Kingdom: a comparative analysis', *Economic History Review*, 2nd series, XXXIII, 3 (1980).

11 Myth and rationality in management decision-making: the evolution of American railroad product costing, 1870–1970

G. L. THOMPSON

In *The Visible Hand* Alfred D. Chandler, Jr explains the rise of big business institutions as rational responses to production bottlenecks in industries with scale economies. Cost accounting, for example, was gradually developed by railroad managers in the period between 1850 and 1880. The technique evolved as an economically rational response to co-ordination challenges posed by the railroad industry's unprecedented scale of geographic dispersion and technical complexity.[1] As other large-scale industries later arose, they borrowed and refined the technique.

While elegant, Chandler's rationality theory raises questions for researchers into the behaviour of firms. For example, my research into the passenger strategy of important American railroads between 1910 and 1960 shows that managements could not predict the marginal cost consequences of their passenger decisions. Acting under a misconceived paradigm of railroad economics, they kept no cost figures on individual passenger trains. Their *ad hoc* calculations of the marginal costs of possible actions usually understated the cost consequences of their actions.

How can such ignorance be reconciled with the depiction of railroad management as the innovator of modern cost accounting? In this article I seek to reconcile the findings of my own research with Chandler's theory of scale-economy rationality. I first examine what Chandler means by cost accounting and contrast his view with product cost accounting. I then set forth the paradigm of railroad economics under which management operated and which made product cost accounting seem irrelevant. Following this, I summarise a hundred years of evidence pointing to the falseness of the traditional paradigm. It is mythology. Finally, I offer reasons for the long persistence of such mythology, which, rather than rationality, provided the foundations of railroad management behaviour.

Chandler's argument that a type of cost accounting arose early in the railroad industry and greatly influenced management is beyond dispute. From early in its history the industry posted labour and material expenses into various accounts for each of their operating divisions. Separate accounts existed for such categories as maintaining locomotives, operating signals or replacing rails on mainline tracks. From 1907 such accounts were kept in accordance with the uniform system of accounts promulgated by the Interstate Commerce Commission (ICC).[2]

Despite criticisms of such accounting methods, they and their attendant management styles produced significant productivity gains for the railroad industry in its earlier years.[3] Between 1839 and 1910 American railroad productivity advanced at the average rate of 3.5 per cent per year, according to a detailed study made by Albert Fishlow. This was very much faster than the 1.3 per cent rate of productivity growth for the economy as a whole. According to Fishlow, productivity growth resulted from ever-longer trains composed of ever-larger cars, more heavily loaded and pulled by ever-larger and more efficient locomotives. The use of cheap steel rails more than any other technological innovation spurred such growth in scale.[4]

Since 1910 railroad strategy has continued unchanged, but the productivity results have been insufficient for the industry to keep up with national productivity. At the end of the 1920s the Harvard economist William Cunningham reported that railroad productivity improvement continued to come from operating longer, larger and heavier trains. He also reported that roadbed and rolling stock investments facilitating such developments no longer earnt a competitive return,[5] perhaps because productivity brought about by such measures failed to match that in manufacturing. Fishlow reports that between 1909 and 1953 railroad productivity advanced at an average of 2.7 per cent per year, but this rate was slower than the rate of productivity growth for the American economy.[6]

The failure of railroad managers to know the costs of the various services they provided, and, hence, whether or not they were profitable, may have contributed to systems profitability difficulties after 1910. It was one thing to know how much it cost on average to maintain a locomotive of a certain class travelling 1 mile. It was quite another to know whether one additional passenger or one additional item of freight moving 1 mile (known as passenger miles and ton miles) added more to the railroad's costs than to its revenues. While managers could estimate roughly the average cost of a passenger mile or ton mile, they had no idea of marginal costs in different circumstances.

Recent management scholars distinguish between the two types of costing activities. According to Robert Kaplan, Robin Cooper and Thomas Johnson, cost accounting applies only to the activity of analysing and controlling operations, which American industry has done well for years. Product costing, on the other hand, refers to the practice of estimating the cost of final products sold in the market-place, a practice that American business has long ignored or misused.[7] Chandler refers to cost accounting, which railroads did indeed pioneer in the several decades following 1840. I refer to product cost accounting, which railroads are just now beginning to master.[8]

Almost from the industry's inception railroad leaders perceived the economics of railroad technology in a way that militated against progress in product costing. Stephen Salsbury's history of the Western Railroad shows that shortly after it opened a through service in 1841 and carried light traffic, its fixed charges exceeded operating costs.[9] Such experiences were common, and fostered the evolution of a paradigm holding that most railroad costs are fixed or constant and that railroad physical plants can accommodate large traffic increases at little additional cost. In the 1850s railroad financial analyst Henry Varnum Poor estimated that 75 per cent of total annual railroad costs were constant. Constant costs included not only interest on debt but also items such as leases, office expenses, and the maintenance of track and rolling stock.[10] Governed by this paradigm, managers were not interested in knowing how much moving a certain amount of traffic 1 mile cost the railroad. Railroad rate-making became the art of arbitrarily assigning what were thought to be constant costs to traffic based on its ability to pay.

Similar views of railroad product costing prevailed well into the twentieth century. In 1900 J. Shirley Eaton, the statistician for the Lehigh Valley Railroad, wrote that variable costs amounted to 5–30 per cent of total annual costs.[11] Emory Johnson, a leading academic expert in railroad transport at the University of Pennsylvania, wrote that 75 per cent of total annual railroad cost was constant in the 1912 edition of his textbook on railroad economics.[12] Isaiah Leo Sharfman's history of the Interstate Commerce Commission published in the early to mid-1930s notes that variable costs were only about one-third of total annual costs.[13] *Fortune* summarised such views in 1936: 'The economics of railroading are exceedingly simple. . . . The gross cost of a train-mile varies almost imperceptibly whether the locomotive is pulling five cars or a hundred.' *Fortune* observed that gross revenues were related to the number of cars, so more cars were better; more cars meant more revenue at negligible cost increase.[14] Albro Martin applied this principle to passenger service in general when he stated that passenger service was viewed by railroad management 'as a high fixed-cost operation in which one more passenger can almost always be carried at nearly clear profit'.[15]

Such a paradigm of railroad economics was the foundation of economically rational American railroad strategy according to important business and institutional historians. It is the basis of Chandler's and Martin's explanations of railroad behaviour.[16] It prefaces Ari and Olive Hoogenboom's history of the Interstate Commerce Commission.[17] Stephen Skowronek places it at the foundation of railroad behaviour in the context of national state building.[18]

What can we conclude if such an important determinant of railroad strategy turns out to be misconceived? Might not the strategy resting on it be irrational? We will now examine over one hundred years of accumulating evidence that points in the direction that the paradigm is, in fact, wrong.

In the 1870s Albert Fink, one of the nation's more respected railroad managers, systematically studied the variation of individual railroad cost accounts in relation to different traffic volumes. He concluded that adding traffic, whether by adding cars to trains or by running more trains, significantly increased most railroad cost accounts, a conclusion corroborated in the 1890s by another southern railway manager, T. M. R. Talcott.[19] Formal statistical corroboration that most railroad costs are variable came from a Ph.D. graduate in economics in 1907. In that year Max O. Lorenz, who later became Chief Statistician for the ICC, published his findings in the *Quarterly Journal of Economics*.[20] Almost a decade later he presented fresh research showing that for mainline traffic densities, almost 100 per cent of costs were variable. This means that if one took the total fixed and operating costs for running a railroad for one year and divided it by the total number of ton miles (1 gross ton moving 1 mile)[21] operated that year, the resulting cost per ton mile would accurately reflect the added or marginal cost of carrying one additional ton 1 mile.[22] Working with Lorenz in the 1920s, Southern Pacific statistician Clarence Day was the first to use regression analysis relating the magnitude of individual railroad cost accounts to traffic volume to ascertain how much of each account was fixed and how much was variable. Day concluded that about 60 per cent of all costs were variable.[23]

Recent econometric work suggests that marginal costs were higher than even the analysts at the end of the 1930s calculated. This revision began with the mid-1950s' work of John Meyer and a group of Harvard economists conducted for the Aeronautical Research Foundation. The Association of American Railroads retained the foundation to study the railroad passenger deficit in 1956 and 1957.[24] The resulting study regressed railroad cost accounts on explanatory variables depicting the volumes of both freight and passenger traffic. The regressions showed how much of each account was constant, and how much each account fluctuated as either freight or passenger volumes changed. Later econometrics work by academic economists such as Theodore Keeler, Anne Friedlaender and Richard Spady used refined methods but came to generally the same conclusions as did Meyer *et al.* during the late 1950s, as well as Lorenz during the first decade of the century.[25]

As Lorenz did in 1916, this body of work concluded that almost 100 per cent of costs are marginal at mainline traffic densities. Once traffic has built up to rather minimal mainline densities, about six or seven mainline goods trains a day in total, still more traffic taxes the resources of the railroad heavily. Longer and more trains require more cars and locomotives, producing greater motive power and car maintenance costs, and more wear and tear on existing facilities. At the same time, they require more yard and terminal facilities, more passing sidings and larger workshops. Such additional facilities and equipment must be operated and maintained. Costs from such sources rise about as fast as traffic, such that whether six or twenty trains a day are operated unit costs are about the same. Moreover, the process is reversible. At any given time, large railroads have facilities in need of replacement. At times of traffic reduction, replacement can be forgone. Yards, sidings, second tracks, roundhouses and workshops also can be operated and maintained at reduced capacity. Some such facilities and rolling stock can be moth-balled altogether, thereby almost completely eliminating their ongoing expenses, or they can be abandoned. Even some debt interest can be reduced. A sizeable part of debt is for rolling stock, which can be sold or returned to the lender. The ability of a railroad to reduce operating expenses and adjust a supposedly fixed plant in the face of traffic reduction is far greater than is commonly thought, as the Great Depression showed.

What is true today about the cost variability of railroad technology was probably true 130 years ago. Noting that the per mile capital charges were higher for the Boston & Worcester, a flat land railroad, than for the Western, a mountainous railroad, Salsbury points out that heavier traffic volumes dictated far heavier capital facilities and costs for even the earliest railroads.[26] The popular notion of a railroad line having unlimited capacity and great scale economies once it is put down is myth.

Kaplan documents in his Union Pacific case study that product costing, and consequently knowledge that marginal costs are far higher than in the traditional paradigm, is now gaining acceptance with railroad management. Why has acceptance taken so long? Kaplan speculates that regulation of rates previously made product costing unnecessary.[27] Another argument is that in the pre-computer era costing of the thousands of individual products moved by railroads between thousands of pairs of stations was too complex a task. I find such arguments unpersuasive. First of all, some railroad managers believed product costing was highly important for proper management under regulation, as the following quote from Talcott in the preface to his cost study published in 1904 illustrates:

> There is a strong and abiding impression in some quarters that railway companies charge high rates on local freight traffic to make up for losses on competitive traffic; and how are we to say that such is not the case if we do not know the cost of doing either local or competitive business? Railways will never be able to make proper defence against this and other charges until they can show with some degree of accuracy what each and every class of business does cost them.[28]

As special counsel to the ICC, Louis Brandeis echoed these sentiments in 1914, and added that railroad inefficiencies arose from product cost ignorance:

> A most surprising difference exists in respect to cost accounting between railroading and manufacturing. Leading American manufacturers know accurately to-day the cost of every one of the numerous articles made and sold by them. Railroads which make and sell a most varied transportation service do not know the cost of any of the services which they furnish. Only a few of the railroads undertake to separate even the cost of freight and passenger service in the aggregate; and among these few there is nothing approaching a consensus of opinion as to the proper basis for such separation. ... In manufacturing accurate cost accounting was found to be a condition precedent to high operating efficiency; and it was found even more essential as a means of insuring the concern against engaging in unremunerative business. What was thus found to be true in manufacturing is equally true in railroading. Cost of service should not, perhaps, determine the reasonableness of a rate; but it is clear that without knowledge of the cost of a particular service it is impossible for railroad officials to protect the company's revenues against unremunerative rates. Carriers' legitimate revenues can not be conserved unless the rate maker has a reasonably accurate knowledge whether a particular service is rendered at a profit or at a loss. In the absence of such knowledge, the traffic manager's success or failure is tested by the tonnage moved instead of profit earned. No adequate explanation can be found for the multitudinous instances of unremunerative rates and practices prevailing on our railroads hereinafter referred to, except lack of knowledge on the part of managers of the disastrous financial result of these rates and practices.[29]

Another explanation for the slow evolution of railroad product costing is that techniques did not exist. But the techniques used today; that is, analysis of individual accounts to determine variability as different types of traffic vary, are essentially the techniques used by Fink in the 1870s and Talcott in the 1890s. That a large staff of accountants would have been necessary in the pre-computer age to analyse product costs does not suggest that product cost accounting would have been uneconomic.

Available evidence suggests other explanations. In his indirect reply to Brandeis on behalf of the eastern carriers, the Pennsylvania Railroad's George Stuart Patterson admitted that railroad leaders were ignorant of product costs. He implicitly justified such ignorance with the argument that cost-based rates were potentially revolutionary; that is, to avoid revolution was to maintain cost ignorance:

> One word as to the cost of service theory which seems to have been developed to some extent in the course of the argument. I do not think I have to argue that proposition to the Commission. The Commission has repeatedly said that rates cannot possibly be made upon the cost of service. There seems to be some idea in this country that

when you say that rates are made upon the principle of what the traffic will bear, that is a principle of extortion. It is not a principle of extortion. It is a principle of moderation. It is the principle of tempering the wind to the shorn lamb. It is the principle upon which a lawyer charges his client. It is the principle upon which a doctor charges his patient. And if that principle of making rates is disturbed in this country, it will revolutionize every freight tariff that there is; it will restrict markets; it will mean that long lines can not meet the competition of the short lines; and as the Commission itself said in the West Virginia case, it will revolutionize absolutely the transportation rates of this country.[30]

This quotation is fascinating, because it shows that it was not the ICC but the railroad industry that mandated itself to cross-subsidise inefficient services (tempering the wind to the shorn lamb), to protect inefficient markets from competition (not restrict markets) and to protect circuitous, inefficient lines (long lines). As Martin shows, the industry reached such a position through a decades-long effort to find peace among its own members and, more importantly, with the collective body of its users.[31] It vehemently resisted efforts that would undo such peace. Nothing was potentially more revolutionary than common knowledge of marginal costs.

A somewhat related railroad fear about product cost accounting was a potential weakening of managerial power if marginal cost knowledge became available to shippers. In 1935 the California Legislature placed for-hire trucks under the jurisdiction of the California Railroad Commission.[32] The legislation directed that trucking rates be based upon the marginal costs of truck operations. Truckers then demanded that a similar standard be applied to rail rates, and pressured the California Railroad Commission to undertake a study of railroad marginal costs.

Acceding to their demands, the Railroad Commission hired University of Southern California transport economist Dr Ford K. Edwards to direct the study. Edwards expanded on the research that Southern Pacific's Professor Clarence Day had previously developed and made use of the railroad's Sacramento workshops for engineering experiments on cost variability.[33]

Unfortunately, the study did not get beyond the stage of developing a general formula for finding the cost of operating different trains, a necessary first step in product costing. According to the Greyhound executive Cloyd Kimball, Southern Pacific first supported and then killed the study. Kimball was in a position to know. He had been one of Edwards's students and after graduation went to work for California's trucking industry. When Edwards went to the Railroad Commission, Kimball followed and participated in the study. According to Kimball, Southern Pacific came to fear the potential political repercussions of the study. At the time most railroad rates still bore no relation to the cost of providing the service. Southern Pacific feared that accurate cost knowledge would motivate those shippers who were paying rates far above the cost of their service to exert political pressure to have their rates legislated downwards. On the other hand, those shippers who were paying rates far below the cost of their service would exert political pressure to have their rates frozen in place. The cold light of truth would force the railroad into bankruptcy, according to this reasoning.[34] Ladd found that as late as 1956 many American railroads resisted the development of accurate passenger train costs for fear that such information could be misused in the wrong hands.[35]

Such anecdotes suggest that railroad management had an interest in not developing accurate product costing. Not only was there an extraordinarily strong corporate culture centred on value-of-product pricing and related industry relations with government, shippers and society at large. Railroad executives also feared change for possible unknown political repercussions. Darkness rather than light kept power in the hands of railroad officers rather than the public.

In examining perhaps the most basic of railroad leader assumptions, that concerning the cost behaviour of railroad technology, this article sides with Thomas C. Cochran's view of rationality in business life. As Chandler later did, Cochran appealed to Weberian theory but was more sceptical about economic rationality as an explanation for the development of railroad managerial practices. Chandler points to dramatic institutional and accounting innovations as rational responses made by railroad managers to relieve production bottlenecks; Cochran sees more fallible human groups behaving in restricted ways dictated by their social environments. Drawing on the darker side of Weberian bureaucratic theory, he sees weaknesses as well as strengths within the solutions organisations developed, a position that he well brought out in his analysis of the development of railroad managerial practices during the nineteenth century.[36]

In examining the support for and the ultimate deposing of the traditional railroad cost paradigm, this article supports Cochran. It shows that for a remarkably long period railroad management adhered to an obsolete view of railroad economics, in spite of abundant evidence to the contrary. The explanation for this long lag is hypothetical, but most likely relates to the social context in which the private American railroad industry found itself in the late-nineteenth century. Once a rate structure was assembled around the initial conception of railroad economics being characterised by high constant costs, subsequent change in the rate structure was next to impossible. Given such a rate structure, there was little incentive for product cost analysis, whose only promise was in upsetting the apple cart of railroad relations with society. In the long run, railroad product cost practices undoubtedly fostered the expansion of wasteful parts of the domestic economy to the detriment of more productive sectors. Such could be said for the railroad industry itself.

Notes

The following abbreviations appear in the notes:

FPC U.S. ICC, *Five Percent Case* (1915), Docket 5860. Record Group 134, Washington National Records Center; decision is reported in 31 ICC 351.

SFC California, Railroad Commission, *Santa Fe Case* (1938), Applications 20710 et al., collection of William Meyer and Jim Seal, Anaheim Hills, Calif. Decision is reported in 41 RCC 239.

1 A. D. Chandler, Jr, *The Visible Hand: the Managerial Revolution in American Business* (Cambridge, Mass., 1977), pp. 109, 115–17, 267–8, 277–9, 464–5.

2 Chandler, *The Visible Hand*, pp. 115–17.

3 S. Salsbury, *No Way to Run a Railroad: the Untold Story of the Penn Central Crisis* (New York, 1982), pp. 10–11, 13, 35, 50–5, 189.

4 A. Fishlow, 'Productivity and technological change in the railroad sector, 1840–1910', *Output, Employment and Productivity in the United States after 1800* (New York, 1966), pp. 583–646.

5 W. J. Cunningham, 'Transportation, Part I. —Railways', in H. Hoover (ed.), *Recent Economic Changes: Report of the Committee on Recent Economic Changes of the President's Conference on Unemployment* (New York, 1929), pp. 255–308.

6 Fishlow, 'Productivity and technological change', pp. 626–30.

7 R. Cooper and R. S. Kaplan, 'How cost accounting distorts product costs', *Management Accounting*, LXIX (April 1988), pp. 20–7; H. T. Johnson, 'Activity-based information: a blueprint for world-class management accounting', *Management Accounting*, LXIX (June 1988), pp. 23–30.

8 R. S. Kaplan, 'Union Pacific, Introduction', 'Union Pacific (A)', and 'Union Pacific (B)', *Case Studies* (Boston, 1985).

9 S. Salsbury, *The State, the Investor and the Railroad* (Cambridge, Mass., 1967), p. 226.

10 A. D. Chandler, Jr, *Henry Varnum Poor* (Cambridge, Mass., 1956), pp.118–19. Fixed costs included interest on debt and leases of other railroads. Review of *Poor's Manual of Railroads* and the ICC's annual *Statistics of Railroads in the United States* shows that such fixed charges amounted to 10–20 per cent of total expenses in the early 1870s, on the eve of the First World War, and in the 1920s and 1930s. They constituted a higher percentage of total costs (generally 25–35 per cent) during the 1880s period of extensive new construction under conditions of high interest rates and diluted traffic.

11 J. S. Eaton, Statistician of the Lehigh Valley Railroad, *Railroad Operations: How to Know Them From a Study of the Accounts and Statistics* (New York, 1900), pp. 282–4.

12 E. R. Johnson, *American Railway Transportation* (New York, 1912), p. 222.

13 I. L. Sharfman, *The Interstate Commerce Commission: a Study in Administrative Law and Procedure* (New York, 1931–37), IIIB, pp. 316–17.

14 'Pennsylvania railroad', *Fortune*, XIII (1936), p. 70.

15 A. Martin, *Enterprise Denied: Origins of the Decline of American Railroads, 1897–1917* (New York, 1971), p. 25.

16 Chandler, *The Visible Hand*, p. 134; Chandler, *Henry Varnum Poor*, pp. 112, 152; Martin, *Enterprise Denied*, pp. 25–6, 43.

17 A. Hoogenboom and O. Hoogenboom, *A History of the ICC: from Panacea to Palliative* (New York, 1976), p. 2.

18 S. Skowronek, *Building A New American State: the Expansion of National Administrative Capacities 1877–1920* (Cambridge, 1982), p. 142.

19 Chandler, *The Visible Hand*, pp.116–19; T. M. R. Talcott, *Transportation by Rail: an Analysis of the Maintenance and Operation of Railroads* (Richmond, Va., 1904), pp. 22–3, 51–2.

20 M. O. Lorenz, 'Constant and variable railroad expenditures and the distance tariff', *Quarterly Journal of Economics*, XXI (1907), pp. 283–98. Lorenz joined the Commission in 1911, becoming Statistician in 1917, a position he held until 1944. See *National Cyclopedia of American Biography*, XLVII (New York, 1965), p. 490.

21 With passenger miles converted to ton miles.

22 M. O. Lorenz, 'Cost and value of service in railroad rate-making', *Quarterly Journal of Economics*, XXX (1916), pp. 205–32.

23 129 ICC 17; 165 ICC 373–392, 410; SFC, 1938, transcript, pp. 11651–2, 11687, 11703, 11718–20; F. K. Edwards, *Study of Rail Cost Finding for Rate Making Purposes, Case No. 4402* (San Francisco, 1938), pp. 138–46, 182–5; C. Kimball, interview with Thompson, St Helena, Calif., 19 May 1985; D. R. Ladd, *Cost Data for Management of Railroad Passenger Service* (Boston, 1957), pp. 39–40, 74, 93–7, 113–15, 120, 131, 142).

24 J. R. Meyer, M. J. Peck, J. Stenason and C. Zwick, *The Economics of Competition in the Transportation Industries* (Cambridge, Mass., 1960).

25 T. Keeler, *Railroads, Freight and Public Policy* (Washington, DC, 1983), pp. 50–3, 153–61; A. F. Friedlaender and R. H. Spady, *Freight Transport Regulation: Equity, Efficiency, and Competition in the Rail and Trucking Industries* (Cambridge, Mass., 1981), pp. 23, 28–35, 217–34.

26 Salsbury, *The State, the Investor and the Railroad*, pp. 109–10, 180, 266, 277.

27 Kaplan, 'Union Pacific'.

28 Talcott, *Transportation by Rail*, p. 17.

29 FPC, *Brief of Louis Brandeis* (1914), pp. 103–5.

30 FPC, 1915, transcript, pp. 22585–8.

31 Martin, *Enterprise Denied*.

32 Edwards, *Study of Rail Cost Finding*, covering letter.
33 Ibid., pp. 35–9; 141–3; 171; Kimball, interview with Thompson (1985).
34 Ibid.
35 Ladd, *Cost Data*, pp. 40–41.
36 T. C. Cochran, *Railroad Leaders, 1845–1890* (New York, 1965), Chapter 6 and pp. 126, 135, 147, 150.

Acknowledgements

I am indebted to the George Krambles Foundation, the National Endowment for the Humanities, the Andrew W. Mellon Foundation and the Hagley Museum and Library for supporting the research that produced this article. It also benefited from thoughtful comments made on earlier drafts by Alfred D. Chandler Jr, Thomas C. Cochran, Brian Gratton, Ed Perkins, Glenn Porter, Maury Klein, Stephen Salsbury and two anonymous referees. I am most grateful for their criticisms and suggestions. Errors and omissions remain my responsibility.